04-SW-4

IMPROVED PROTOCOL FOR CLASSIFICATION AND ANALYSIS OF STORMWATER-BORNE SOLIDS

by:

Dr. Larry A. Roesner (PI)
Dr. Amy Pruden (Co-PI)
Elizabeth M. Kidner
Colorado State University

2007

WERF

The Water Environment Research Foundation, a not-for-profit organization, funds and manages water quality research for its subscribers through a diverse public-private partnership between municipal utilities, corporations, academia, industry, and the federal government. WERF subscribers include municipal and regional water and wastewater utilities, industrial corporations, environmental engineering firms, and others that share a commitment to cost-effective water quality solutions. WERF is dedicated to advancing science and technology addressing water quality issues as they impact water resources, the atmosphere, the lands, and quality of life.

For more information, contact:
Water Environment Research Foundation
635 Slaters Lane, Suite 300
Alexandria, VA 22314-1177
Tel: (703) 684-2470
Fax: (703) 299-0742
www.werf.org
werf@werf.org

This report was co-published by the following organizations. For non-subscriber sales information, contact:

IWA Publishing
Alliance House, 12 Caxton Street
London SW1H 0QS, United Kingdom
Tel: +44 (0) 20 7654 5500
Fax: +44 (0) 20 7654 5555
www.iwapublishing.com
publications@iwap.co.uk

Library of Congress Catalog Card Number: 2006938719
IWAP ISBN: 978-1-84339-776-2

This report was prepared by the organization(s) named below as an account of work sponsored by the Water Environment Research Foundation (WERF). Neither WERF, members of WERF, the organization(s) named below, nor any person acting on their behalf: (a) makes any warranty, express or implied, with respect to the use of any information, apparatus, method, or process disclosed in this report or that such use may not infringe on privately owned rights; or (b) assumes any liabilities with respect to the use of, or for damages resulting from the use of, any information, apparatus, method, or process disclosed in this report.

Colorado State University

The research on which this report is based was developed, in part, by the United States Environmental Protection Agency (EPA) through Cooperative Agreement No. CR-831559-01 with the Water Environment Research Foundation (WERF). However, the views expressed in this document are solely those of Colorado State University and neither EPA nor WERF endorses any products or commercial services mentioned in this publication. This report is a publication of WERF, not EPA. Funds awarded under the Cooperative Agreement cited above were not used for editorial services, reproduction, printing, or distribution.

This document was reviewed by a panel of independent experts selected by WERF. Mention of trade names or commercial products does not constitute WERF nor EPA endorsement or recommendations for use. Similarly, omission of products or trade names indicates nothing concerning WERF's or EPA's positions regarding product effectiveness or applicability.

ACKNOWLEDGMENTS

The principal investigators express their thanks to the members of the Project Subcommittee (PSC) for their enthusiastic support, helpful review, and constructive inputs to this research product. The researchers wish to thank also the peer reviewers for their review and constructive comments. Thanks go also to the following individuals who made special contributions to this project: Dr. Richard Ashley, who served as a peer reviewer of the report and brought to our attention a large body of European literature relevant this research; Dr. Michael Barrett for his contribution to regulatory issues regarding stormwater solids analysis found in Chapter 10.0; Dr. Michael Stenstrom who provided data from his laboratory research on the relationship of measured suspended solids concentrations to sample mixing speed, and to James Lenhart for providing data from his laboratory experiments on the relationship of measured suspended solids concentrations to sample settling time. Special thanks go to Elizabeth Kidner who served as project manager and used this research as her Masters Thesis. Liz's ability to work independently, prepare draft deliverables, and coordinate with the PSC and the peer reviewers contributed greatly to the quality and product of this research. The Harold H. Short Endowed Chair Position in Colorado State University's Civil and Environmental Engineering Department is acknowledged for supporting the in-kind services provided by CSU on the project.

Report Preparation

Principal Investigators:
Larry A. Roesner, Ph.D., P.E. (PI)
Colorado State University - Department of Civil and Environmental Engineering
Harold H. Short Urban Water Infrastructure Laboratory

Amy Pruden, Ph.D. (Co-PI)
Colorado State University – Department of Civil and Environmental Engineering

Project Team:
Elizabeth M. Kidner, EIT, Graduate Research Assistant
Colorado State University – Department of Civil and Environmental Engineering

Project Subcommittee

Ben Urbonas, P.E., D.WRE
Urban Drainage and Flood Control District

Michael Bloom, P.E., C.F.M, D.E.E
Post, Buckley, Schuh & Jernigan (PBS&J)

Richard Field, P.E., D.WRE
U.S. Environmental Protection Agency

Qizhong Guo, Ph.D., P.E.
Rutgers University

James H. Lenhart, P.E.
CONTECH Stormwater Solutions Inc.

Darrin Peine, QEP
City of Charlotte

Water Environment Research Foundation Staff

Director of Research: Daniel M. Woltering, Ph.D.
Senior Program Director: Jeff C. Moeller, P.E.

Peer Reviewers

Richard Ashley, Ph.D.
Department of Civil and Structural Engineering University of Sheffield, UK

Michael Barrett, Ph.D., P.E.
Center for Research in Water Resources *University of Texas*

Betty Rushton, Ph.D.
Environmental Scientist, Southwest Florida Water Management District

Linda D. Pechacek, P.E.
LDP Professional Services

Michael K. Stenstrom, Ph.D., P.E., D.E.E
Civil and Environmental Engineering Department UCLA

ABSTRACT AND BENEFITS

Abstract:

A large portion of water quality impaired waterways are located in or near urban areas and are adversely influenced by stormwater-borne solids. The solids have negative impacts on receiving water systems including loss of aquatic habitat, channel instability, and the transport of harmful pollutants potentially hazardous to human and ecosystem health. The current methods for sampling, handling, and analyzing stormwater solids do not lead to a good understanding of these effects on receiving waters. The purpose of the study was to develop a consistent classification of stormwater solids and to recommend modifications to the analytical methods for determining stormwater-borne solids in order to improve assessment and monitoring protocols.

Currently accepted practices for characterizing stormwater-borne solids are critically analyzed and revised. Common definitions and standardized monitoring procedures are recommended in this report to aid in understanding solid impacts and selection of stormwater best management practices. Stormwater solids should first be classified based on size into dissolved, fine, coarse and gross solids. These solids can further be classified as settleable or suspended by allowing a settling time in the analytical procedure for Total Suspended Solids. Obtaining a representative sample in the field is one of the biggest challenges in characterizing stormwater-borne solids because of temporal, geographic, and spatial variations. An outline for developing a monitoring plan for fine solids and gross solids is described.

Benefits:

- Provides a detailed literature review and synthesis of the existing sample collection, handling, and analysis methods.
- Develops a consistent classification system that defines the major classes of stormwater solids including solids known as gross solids.
- Recommends a draft protocol addressing sample collection, handling, and analysis that can be used to monitor stormwater solids.

Keywords: Total suspended solids, suspended sediment concentration, gross solids, stormwater monitoring, water quality.

TABLE OF CONTENTS

LIST OF TABLES

LIST OF FIGURES

LIST OF ACRONYMS

AMAFCA	Albuquerque Metropolitan Arroyo Flood Control Authority
APHA	American Public Health Association
ASTM	American Society for Testing and Materials
AWWA	American Water Works Association
BMP	Best Management Practice
BOD	Biochemical Oxygen Demand
DO	Dissolved Oxygen
EMC	Event Mean Concentration
FHWA	Federal Highway Administration
MEP	Maximum Extent Practicable
MS4	Municipal Separate Storm Sewer System
NPDES	National Pollutant Discharge Elimination System
NRDC	Natural Resources Defense Council
NURP	National Urban Runoff Program
PAH	Polycyclic Aromatic Hydrocarbon
PSD	Particle Size Distribution
SM	Standard Methods
SSC	Suspended Sediment Concentration
TDS	Total Dissolved Solids
TPH	Total Petroleum Hydrocarbons
TS	Total Solids
TSS	Total Suspended Solids
U.S. EPA	United States Environmental Protection Agency
USGS	United States Geological Survey
UWRRC	Urban Water Resources Research Council

EXECUTIVE SUMMARY

ES.1 Introduction

Stormwater-borne solids have been documented to have negative impacts on receiving water systems and can lead to loss of aquatic habitat, cause channel instability, and transport harmful pollutants. Waterways located in or near urban development are adversely influenced by solids in stormwater runoff. Total suspended solids is commonly reported in stormwater monitoring, but may not accurately represent the true character of the solids in the water. Moreover, vagueness in the Standard Methods laboratory protocol means that not all TSS is measured using exactly the same protocol. The establishment of consistent solids definitions, sampling, and analysis techniques for accurately characterizing stormwater solids and associated pollutants is currently a critical need for effective stormwater management.

Recognition of the potential adverse impacts of stormwater runoff pollutants led to stormwater permitting under the National Pollutant Discharge Elimination System (NPDES) programs and to stricter regulations being applied to new urban development. While the potential impacts of various stormwater-borne solids are recognized and regulatory action is being taken, no common definitions or standardized monitoring procedure exists for them. This has hindered efforts to characterize stormwater-borne solids and to better understand the impacts of these materials on receiving water systems.

The objective of the literature review is to summarize the current state of stormwater solids characterization and sampling techniques, identify advantages and disadvantages of identified techniques, and highlight existing concerns surrounding stormwater monitoring. The review summarizes the research findings regarding Total Suspended Solids, Suspended Sediment Concentration, and Gross Solids. Based on the literature review, a classification system is proposed that defines the major classes of stormwater-borne solids. Finally, protocols addressing sample collection and analysis have been developed that can be used to monitor the new classes of stormwater-borne solids.

ES.2 Solids Definition

One of the fundamental obstacles in stormwater management is the lack of consistent definitions of stormwater solids. This report explores a classification system that has been developed for stormwater solids based on size taking into account sampling techniques, ecologic impact, and potential treatment. Furthermore, the various solids classifications can be characterized analytically as suspended, settleable, volatile, or non-volatile.

It is proposed herein that *gross solids* be defined as solids greater than 5 mm including litter, debris, and coarse sediment. Litter includes human derived solids such as trash, plastic, clothes/fabric, Styrofoam, and glass. Debris includes organic matter such as twigs, grass, and leaves. *Coarse solids* are defined as solids between 75 μm and 5 mm. In general, this includes sand size sediment and larger. Coarse solids have been shown to carry a large amount of metals and other toxic compounds into the waterways. In addition, coarse solids tend to settle and infill habitat areas, smother benthic organisms and fish eggs, and change bedforms that are necessary for aquatic habitat. *Fine solids* are defined as solids less than 75 μm and greater than 2 μm, including silt and clay. Fine solids are attributed to transporting harmful constituents into receiving waters, increasing suspended solids and turbidity, and reducing the numbers of sensitive organisms. *Dissolved solids* are the solids that pass through a 2 micron filter and are usually not treated using traditional Best Management Practices (BMP) that rely on settling.

The solids can further be described as settleable, suspended, volatile and non-volatile based on analytical characteristics. Evaluating these characteristics offers more detailed understanding of how the solids will be transported and the ultimate fate of the solids in the stormwater runoff. In addition, understanding the settling characteristics can offer information on the proper treatment processes to remove solids before they enter the natural waterway. Although this report focuses on stormwater-born solids, the method of classifying solids has application to streams, sanitary sewer overflows, and combined sewer overflows.

ES.3 Solids Analytical Protocol

Many municipalities and stormwater management programs have adopted criteria for expressing the "quality" of stormwater and the effectiveness of their control programs in terms of Total Suspended Solids (TSS). Error is introduced in the TSS laboratory analysis method due to differences in mixing speeds and methods, subsample location, and equipment used. Suspended Sediment Concentration (SSC) analyzes the entire sample, reducing the error associated with subsampling. However, SSC may not be appropriate for determining the types of suspended solids that are important for determining environmental impact because the SCC analysis will cause the focus for solids removal (especially when "percent removal" is the BMP target) to be on the heavier solids, whereas the problem may be the finer solids that have sorbed nutrients, heavy metals and toxic materials and therefore should be the target for removal.

It is recommended to that the current procedure for analyzing Total Suspended Solids revised at a minimum as follows. For standardization, the sample should be filtered over a No. 4 sieve in the U.S. standard sieve sizes to separate gross solids from the sample. The sample should then be mixed at 600 rpm's using a magnetic stirrer for one minute after which a aliquot is taken from the approximate middle of the sample, mid-way between the side of the sample container and the vortex using a large bore pipette. The subsample should then be filtered through a 2 micron filter (not a 1-2 micron filter as the current protocol permits) and the residue analyzed to determine the TSS concentration. Using this modified procedure will help standardize the TSS method by identifying mixing speeds and aliquot sample location.

But this minimum refinement still does not adequately address the issue of heavier solids in the sample because the aliquot sampling technique of individual investigators will result different amounts of the heavy solids being extracted from the main sample. Thus it is recommended that a period of settling be included in the protocol after mixing and before aliquot subsampling. Two analytical procedures were examined to include a period of settling to separate the rapidly settling solids from the solids that settle more slowly. One method is to take the TSS aliquot after a short settling time following mixing of the sample. This will allow for the heavier settling solids to move toward the bottom of the container and not be included in the suspended solids analysis. The second method is to use a separatory funnel for separation of settleable and suspended solids. This would allow the entire water sample to be analyzed (similar to the SSC analytical procedure) by characterizing the settleable solids separate from the suspended solids. At this point a settling time of 5 minutes is recommended, but further testing should be done to compare TSS determinations as a function of settling time to determine the best settling time to obtain the desired type (basically specific gravity) of solids of interest measured by the TSS test. More experimental work needs to be conducted with the separatory funnel to confirm its efficacy and ability to replicate the SCC determinations of solids content in a sample.

Total Suspended Solids procedure currently lumps all solid material into a single class (i.e. solids less than or equal to a given specific gravity, depending on aliquot sampling technique). Particle size distributions, however, can offer additional information about the solids of concern in the runoff and the potential methods for treatment. But, solids in stormwater are in a dynamic state and change characteristics due to flocculation, degradation, and aggregation; therefore, if particle size distribution is to be measured, it is recommended that the maximum holding time be less than six hours. Analysis on chemical constituents associated with the solids, including metals, PAHs, toxic organics, and nutrients, is also recommended based on the water quality goals.

ES.4 Sampling and Monitoring

Obtaining a representative stormwater sample in the field proves to be extremely difficult because stormwater is so unpredictable. Solids in stormwater runoff are variable in size, concentration, time and location. The irregular intensities of precipitation make it difficult to predict runoff rate, sediment transport, deposition and re-suspension, etc. Landscape practices, spills, construction activities, traffic density, and vehicle washing can drastically influence the runoff characteristics. With these influences, it is not surprising that stormwater characteristics and quality are highly variable from location to location and from storm event to event.

Due to the large variations in stormwater characteristics, it is recommended that monitoring plans be developed specifically for the location and goals of the stormwater management program. There are several existing documents that give guidelines on stormwater sampling and how to obtain a representative sample in the field.

Two very difficult classes of solids to measure accurately are gross solids and course solids. It is recommended gross solids samples be collected (preferably) in a gross solids removal device or by using nets or screens with 5 mm openings. Coarse solids are usually ignored in stormwater sampling because grab samples are usually taken at the water surface and the intake of autosamplers is usually located above the bottom of the channel, and the nozzle size is too small. Therefore, it is recommended to use bed load samplers in addition to grab sampling or autosampling to better sample the particle size distribution transported in the stormwater. The location of sampling points should be judiciously determined to best represent the solids in the stormwater runoff and to meet the goals of the monitoring program.

Accurate reporting is one of the most important aspects of a monitoring program. Detailed reports on field sampling, equipment, sample location, and methods used are necessary to improve research efforts. In addition, detailed reports on analytical methods are required including holding time and temperature, mixing speed, equipment used and filter size.

ES.5 Summary and Recommendation

Currently accepted practices for characterizing stormwater-borne solids need to be critiqued, re-thought, and revised. Stormwater-borne solids include suspended sediment, bed-load, settleable and non-settleable solids, gross solids, as well as organic and other natural material. These solids can have adverse impacts to receiving water systems and can lead to loss of beneficial uses. This report summarizes the current state of stormwater solids characterization and sampling techniques and suggests improved monitoring methods. A standardized classification system is identified based on particle size. Draft protocols for improved analytical protocol and monitoring methods are discussed.

Two analytical methods are identified which include a settling time for coarse solids separation. It is recommended that research and validation be done using these methods. Understanding the settling characteristics of the solids offers a lot of information on transportation, ultimate fate, and treatment options.

It is recommended that further research be done on improving field sampling methods to obtain representative solids samples. Solids are variable in concentration and characteristics vertically as well as horizontally in runoff making it difficult to obtain representative samples in the field. Improving sampling equipment and methods will result in improved stormwater monitoring and the understanding of stormwater solids and their associated impacts.

The more that is learned about the environmental impacts of stormwater solids the more information is required to advance our current monitoring and analytical methods. Linking stormwater best management practices to the defined categories of solids can improve our current management strategies and improve the ecologic impacts of urban development. It is recommended that the solid classification defined within this report be linked back to best management practices to improve current stormwater management strategies.

CHAPTER 1.0

STORMWATER SOLIDS BACKGROUND

1.1 Introduction

The more that is learned about the processes that affect the aquatic environment, the more detail that will be required in describing the water quality parameters that affect it. This is especially true for stormwater-borne solids. Stormwater-borne solids are a major contributor to surface water quality degradation in waterways. A large portion of water quality impaired surface waters is located in or near urbanized areas and stormwater is a major source of the contamination. In particular, the U.S. Environmental Protection Agency (2000) has identified sediment as the most widespread pollutant in U.S. rivers and streams, affecting aquatic habitat, drinking water treatment processes, and recreational uses of rivers, lakes, and estuaries. The establishment of consistent solids definitions, sampling, and measuring techniques for accurately characterizing stormwater solids and associated pollutants is currently a critical need for effective stormwater management. This chapter defines the commonly accepted definitions of stormwater-borne solids, and discusses some typical solids regulations and BMP goals. Chapters 2.0, 3.0, 4.0 and 5.0 comprise a literature review of issues and current practice as they relate to stormwater-borne solids. Chapter 2.0 presents an overview of why proper management of stormwater-borne solids is important with respect to geomorphology, ecology, and water quality; Chapter 3.0 reviews current solids characterization practice. Chapter 4.0 reviews current practices for sample collection and handling while Chapter 5.0 addresses current laboratory protocols used for solids characterization. A synthesis of the literature review is provided in Chapter 6.0. The literature review provides a background for recommending a consistent method for defining stormwater solids in a format that meets the needs for effective stormwater management (Chapter 7.0), followed by a proposed protocol for stormwater solids analysis in Chapter 8.0, and general guidance for sampling in Chapter 9.0. The last Chapter (10.0) discusses regulatory issues related to substituting SSC for TSS as the measure of solids removal from stormwater discharges. Although this report focuses primarily on stormwater-borne solids, the method for classification and protocol for evaluating suspended solids are applicable to solids carried in streams, sanitary sewers, and combined sewers.

1.2 Overview of Stormwater Solids

Total Suspended Solids (TSS) is the hallmark water quality parameter for quantifying the concentration of solids in stormwater, and it is written into the stormwater discharge permits of numerous municipalities without much regard to the actual impacts of TSS on the receiving waters. The inclusion of TSS in stormwater permits is an extension of its use in municipal and industrial wastewater treatment, where it is used to aid in the design of sedimentation tanks for the removal of suspended solids. But its use in stormwater management has been extrapolated to serve as an indicator of many receiving water responses to stormwater discharge including scour and deposition potential, visibility impairment, benthic filter feeder clogging, and organic solids sedimentation rates. However, the analytical protocol is not well suited to give a valid estimation of the real effect that suspended solids have on the processes of interest.

Increasingly the use of TSS as an indicator of stormwater pollution has resulted in erroneous BMP solids removal estimates, inappropriate TMDLs and other water quality regulations that do not really solve the problem from an environmental standpoint. In order to make progress with this issue, it is necessary to have a better more detailed characterization of the solids historically called TSS. However, it is important that the methods of sampling and the analytical protocols be consistent, reproducible, pragmatic, and of reasonable cost so that municipalities and other water quality agencies can understand them and afford to carry out the collection and analysis. This document sets developing consistent definitions of stormwater-borne solids and improving methods monitoring for monitoring them as the primary goal.

Stormwater solids include suspended sediment, dissolved solids, settleable and non-settleable solids, litter, debris, floatables, and coarse solids. These solids can have various negative impacts on the receiving waters and can pose a physical problem affecting geomorphology and ecologic habitats in addition to potentially transporting harmful chemicals from the urban environment into the natural ecosystems. Though the adverse impacts are generally recognized, there is still not a common definition of solids nor a standardized procedure for their sampling and monitoring. This poses a significant impediment to the appropriate management and treatment of stormwater solids.

Standard Methods defines TSS as the portion of total solids (TS) retained by a 1-2 micron filter (APHA, 1998). TS is defined as the sum of TSS and Total Dissolved Solids (TDS) where TDS is determined as the filtrate residue left in a vessel after evaporation of a sample and subsequent drying in an oven at a defined temperature (SM 2540 B (TS) SM 2540 C (TDS), SM 2540 D (TSS)). So, strictly speaking, TSS includes litter, trash, gross solids in both organic and inorganic forms in addition to floating and sinking particles, sand and cohesive soil solids, and organic solids. In an effort to circumvent the problems with using TSS to describe all these types of particulate matter in stormwater, investigators have developed/proposed new descriptors and tests to better compartmentalize these solids. Some examples follow.

Urban Litter is defined by Armitage and Rooseboom (2000a, 2000b) as visible solid waste emanating from the urban environment with an average dimension of greater than about 10 mm and typically comprising: plastics, paper, metals, glass, vegetation, animals, construction material, miscellaneous articles such as clothing, pens and pencils, cigarette butts, old tires, etc. Lloyd and others (2001) define urban litter as manufactured items that can be retained by 6.35-mm mesh, while Kayhanian and others (2005) propose a more simple classification:

- *Non-Biodegradable litter*: litter waste that does not naturally degrade in the environment, such as metals and plastics.
- *Biodegradable*: litter waste that naturally degrades in the environment, such as paper.

Gross pollutants are related to urban litter but have no standard definition either. Allison and others (1998) define gross pollutants as litter and debris greater than 5 mm in size. Rushton and England (2006 Draft Report) define gross pollutants as litter, debris and sediments larger than 75 μm that travel in suspension or bedload, and are subdivided into three types: 1) litter and, in general, human derived trash; 2) debris composed of organic material; and 3) coarse sediments composed of organic products from soil, pavement or building material. Dallmer's (2002) definition is material larger than 2 mm in size, while Shaheen (1975) and Mudgway et al. (1997)

define it as material larger than 2 or 3 mm in size. In Europe, where a primary contributor to gross solids is from combined sewer overflows, gross solids are defined as solids with a specific gravity close to 1.0 and can be captured by a 6 mm screen (Butler et al., 2002; Jefferies and Ashley, 1994).

There are other problems associated with the measurement of gross solids. James (1999) notes four problems: 1) sampling protocols for such measurements require collection of large volumes of stormwater, 2) techniques for the continuous or on-line measurement of the wide range of particle sizes have not been used in stormwater monitoring, 3) commonly used peristaltic automatic sampling equipment does not appear to be capable of collecting representative samples of gross or total solids in stormwater runoff and 4) sampling points are placed above the bed to prevent clogging of automatic sampling equipment therefore excluding bedload from the sample. Floating material is also excluded in sampling because of the placement of sampling equipment. In addition to inconsistent classification schemes and definitions used in research and characterization studies, there are inconsistent regulatory definitions of these types of pollutants as well.

Total Suspended Solids was originally defined for wastewater transport and urban runoff based on a filtration procedure according to Standard Methods (Standard Methods 2540). Some researchers have also allowed a sedimentation time before the test, such as a study by Stahre and Urbonas (1990) which allowed a five minute sedimentation period to allow for a coarse separation before performing the TSS analysis. The suspended solids are the main vector of wet-weather pollution in combined sewers. For example, in and experimental urban catchment with combined sewers 'Le Marais', Chebbo and others (2005) found 60-95% of organic matter, 65-99% of zinc and hydrocarbons, 90-100% of cadmium, copper and lead for the same storm are linked to suspended solids, as defined by the standard tests.

In urban runoff, solids are also often found to be a carrier of other pollutants such as nutrients, toxins, heavy metals and other organic and inorganic materials (Chebbo and Bachoc, 1992; Muthukaruppan et al., 2002). But currently, most water quality monitoring studies are concerned only with pollutant loads, and not with their association with the solids fraction. However, Muthukaruppan and others (2002) and many other researchers report that the particle size distribution of the solids and the pollutant load distribution across different particles sizes are critical in determining the pollutant transport capacity of these solids and the treatability of the runoff. Sansalone and Burcherger (1997) reported that the mass of solids, heavy metals and nutrients such as total phosphorous are associated with particles greater than 150 μm, which represents a diameter size that is not well measured with the current TSS analysis because these larger (heavier) particles tend to settle to the bottom of the sample.

Size and mass density are the main sediment characteristics that determine the movement of solids, whether as suspended, washload or bedload. Hijioka and others (2000) identified two fractions as separated by a size of 45 μm. This limit represents the difference in the dependence of the pollutant loads and the intensity of storm. This value also corresponds to the limit between sand and silt (62 μm), which represents the critical size to differentiate the results for Suspended Sediment Concentration (SSC) and TSS (Gray et al., 2000). If the sample contains a substantial amount of sand size particles stirring or mixing the sample using a magnetic stirrer will rarely produce a subsample that is representative of SSC because of the rapid settling characteristics of sand-size particles compared to those of silt and clay (Gray et al., 2000).

<u>*Suspended Solids vs. Settleable Solids in Suspension*</u> The TSS analytical protocol does not distinguish between settleable and non-settleable solids, or between the "settleability potential" (i.e. density, size, and organic content/flocculation potential) of the solids that pass through the 1-2 micron filter. Yet this fraction is extremely important in assessing the ecologic impact of stormwater discharges to streams and the potential of treating the runoff before discharge. The fundamental difference between the SSC and TSS analytical methods stems from preparation of the sample for subsequent filtering, drying, and weighing. The method for determining SSC requires the filtration and analysis of the entire sample, while the method for TSS requires the withdrawal of an aliquot for filtration and analysis. These methods are discussed in detail in Section 4.1. It has been reported (James, 1999; Gray et al., 2000) that the method for determining TSS may be fundamentally flawed for the analysis of natural water samples, such as stormwater discharges, particularly when sand-size material comprises a substantial percentage of the sediment in the sample. In contrast, the method for determining SSC (Gray, 2002) produces relatively reliable results for samples of natural water, regardless of the amount or percentage of sand-size material in the samples. Additionally, the percentage of sand-size and finer material can be determined as part of the SSC method, but not as part of the TSS method. Gray demonstrates that at similar flow rates, sediment discharge values from SSC data can be more than an order of magnitude larger than those from TSS data (Gray et al., 2000) due primarily to heavier particles that are often missed in the TSS method.

<u>*Settleable Solids*</u> The settleable portion of TSS, which is not reported in most stormwater studies, can settle to the bottom of water bodies and damage invertebrate populations, cause imbalances in stream biota, reduce spawning gravels, remove dissolved oxygen from the water, reduce the pH, reduce conveyance capacities and increase dredging frequencies and costs for treating the solids. Trash and debris in stormwater runoff are more than an aesthetic problem despite being one of the main public concerns in regard to waterway health. Plastic particles, soda rings and Styrofoam particles have been found in the stomachs of seabirds and other marine life (U.S. EPA, 1992; Blight and Burger, 1997).

As discussed above, the USGS has developed a modified analytical protocol called the "Suspended Sediment Concentration" for which they have provided significant evidence to support the conclusion that it provides a more accurate estimate of the solids concentration in the water for geomorphic studies. But the new test does not necessarily indicate ecologic impacts.

1.3 Solids Regulations and BMP Goals

Solids in stormwater have been regulated since the early 1990s after Congress amended the Clean Water Act in 1987 to impose National Pollutant Discharge Elimination System (NPDES) permitting requirements on pollutants in stormwater runoff. These requirements were added to EPA regulations in two phases. The first phase of regulatory action required municipal separate storm sewer system (MS4) operators with populations over 100,000 people to obtain an NPDES permit, develop and implement a Stormwater Management Program to reduce pollutant discharges to the *maximum extent practicable*. The second phase of regulatory changes required all MS4 operators located in U.S. Census defined urbanized areas, regardless of population, to do so. For additional information on these regulatory requirements consult Part 122.26 of Title 40 of the Code of Federal Regulations (40 CFR 122.26).

1.3.1 Goals of BMPs

There are several overarching goals of Best Management Practices (BMPs) including improving hydraulics and water quality. Improving the downstream aquatic environment through erosion control and minimizing environmental risks are principal goals of BMPs. They improve runoff characteristics, reduce erosion and scour, and mitigate flooding. BMPs have many water quality impacts such as reducing downstream pollutant loads and concentrations, removing litter and debris, and reducing toxicity of runoff. Many of these BMP goals can be evaluated using flow and water quality monitoring, performing different laboratory analyses depending on the issues of concern. Although, in some cases, the goals cannot be directly evaluated using water quality monitoring and additional analyses is required.

Several state regulations were reviewed to provide typical examples of current stormwater management programs. These are summarized below.

New Jersey

New Jersey has put into place one of the strictest stormwater regulations in the country. Their stormwater BMP manual outlines their comprehensive stormwater management program requiring the following design standards:

- All projects must be designed so that the peak post-development runoff rate from the 2-, 10-, and 100-year storm event do not exceed 50, 75, and 80% of the predevelopment peak runoff rates respectively.
- Stormwater treatment BMPs must be designed to remove 80% of the TSS load on an annual basis for "major development" projects that create at least 0.25 acres of new or additional impervious surfaces.
- Stormwater management measures must reduce the average annual nutrient load by the maximum extent feasible.
- BMPs must be designed to treat the flow from a water quality storm of 1.25 inches of rain (non-uniformly distributed) in 2 hours
- BMPs must maintain 100% of the average annual preconstruction groundwater recharge volume for the site or the increase of stormwater runoff volume from pre-construction to post-construction for the two-year storm must be infiltrated.

Washington

The state of Washington also provides stringent stormwater regulations. Some of their design criteria are listed below.

- BMPs in Washington must be designed to remove 80% of the TSS load during the peak of the 6-month, 24-hour storm.
- All stormwater treatment devices must be designed so that peak discharges from the 2-, 10-, and 50-year, 24-hour storm do not exceed predevelopment rates
- Devices being evaluated must demonstrate an effluent TSS concentration of 20 mg/l when influent concentrations are less than 100 mg/l, and 80% removal of TSS if the influent concentration is greater than or equal to 100 mg/l, using a silt loam particle size distribution.

Maine

Maine adopted new rules to address stormwater treatment objectives in 2005 with goals of pollutant removal, temperature control, channel protection and flood control. BMPs that are approved for these goals include underdrained soil media filters, wet ponds with an underdrained gravel filter outlet, infiltration practices, and vegetated buffers. They also allow hydrodynamic solids separators, that have been tested and approved by the State, to be used as pretreatment devices upstream of BMPs impacted by sedimentation. The State has moved away from using TSS as a surrogate pollutant in favor of more stringent standards which result in more effective removal of nutrients, metals and other stormwater pollutants.

North Carolina

The North Carolina Department of Environment and Natural Resources (NCDENR) have established stormwater requirements including the following:

- Projects must maintain low densities of impervious area, which ranges between 12-30% depending on the project location.
- Projects must maintain vegetated buffers and transport runoff through vegetated conveyances.
- If the above criteria cannot be met, projects must install structural stormwater BMPs capable of controlling the runoff from the 1.0 or 1.5 inch rain event. BMPs must also remove 85% of the TSS on an annual basis.

California – Trash TMDLs

Los Angeles Regional Water Quality Control Board in 2001 established a requirement on trash total maximum daily loads (TMDL) for the Los Angeles River and Ballona Creek. The trash TMDL establishes a waste load allocation of zero by September 2013. Trash is defined as litter and particles that can be retained by a 5-mm mesh screen, where litter is defined as "all improperly discarded waste material, including, but not limited to, convenience food, beverage, and other product packages or containers constructed of steel, aluminum, glass, paper, plastic, and other natural and synthetic materials, thrown or deposited onto the lands of waters of the state, but not including the properly discarded waste of the primary processing of agriculture, mining, logging, sawmilling or manufacturing…" (Government Code Section 68055.1(g))

New York-New Jersey

New York and New Jersey started a Harbor Estuary Program (HEP) in 1987 with the goal to manage floatable debris in the area. Their goals include eliminating floatables-related beach closures, prevent adverse impacts on commercial and recreational boating and on coastal species from floatable debris. In this study, floatable debris is defined as waterborne waste material that is buoyant. Examples include wood, beach litter, aquatic vegetation, and detritus; street litter (cans, bottles, polystyrene cups, sheet plastic, straws, and paper products); sewage-related wastes (condoms, sanitary napkins, tampon applicators, diaper liners, grease balls, tar balls, and fecal material); fishing gear (nets, floats, traps, and lines); and medical wastes (hypodermic needles, syringes, bandages, red bags, and enema bottles) (NY-NJ HEP, 1996).

The above list of state regulation is by no means an exhaustive list, but it gives an overview of common regulations required at the state level. Many states use the federal NPDES program as a guideline, requiring that stormwater be treated to the "maximum extent practicable (MEP)". A common requirement for construction sites disturbing more than one acre, many industrial sites, and all designated Municipal Separate Storm Sewer Systems (MS4's) is to obtain permit coverage for stormwater discharge.

CHAPTER 2.0

IMPACTS OF STORMWATER SOLIDS

2.1 Introduction

As noted earlier, stormwater-borne solids have detrimental impacts on receiving waters, affecting geomorphology, and aquatic habitats, the chemistry of the water. Bacteria in stormwater also affect recreational activities. The impacts of stormwater solids are variable according to the urban and natural environment to which they are discharged, and also depend on the local climate and weather patterns. However, typical symptoms of impacted urban streams include degradation in geomorphology, water quality, ecology, and biodiversity. Stormwater solids have been shown to change the bed material, increase suspended sediment loads, result in a loss of riparian habitat due to erosion, and change the variability of flow and sediment transport characteristics relative to aquatic life cycles (Roesner and Bledsoe, 2003). It is also common to have increases in erosion and incision (Wolman, 1967; Roberts, 1989; Booth, 1991). An increase in contaminated load and concentrations of harmful constituents is also an observed result of increased solids in the runoff (Osborne and Wiley, 1988; Corbett et al., 1997; Hatt et al., 2004). Finally, there is likely to be an overall decrease in biodiversity, including genetic, species, and community levels (Richter et al., 1997; Chessman and Williams, 1999; Walsh et al., 2004). The following sections describe these adverse impacts more specifically.

Stormwater runoff is attributed to transporting various harmful constituents into the water ways. For example, bacterial contamination from animal waste and organic fertilizers washes into waterways during storm event runoff causing hazards to the health of the public who come in contact with the water through recreational activities. Bacteria and pathogens leaching from litter or medical waste can cause disease as well. Common pathogens found in urban runoff include, but are not limited to, *Salmonella, Pseudomonas aeruginosa, Straphylcoccus aereous, Escherichia coli*, fecal *Streptococci* and *Shigella*. When humans or pets come into contact with pathogens it may cause illness or other discomfort such as ear infections, skin irritations, chills, and gastrointestinal illness (Pitt et al., 2001).

Bacteria and pathogens found in stormwater runoff potentially cause human health problems through exposure to contaminated stormwater by swimming or recreational activities, drinking water contaminated by stormwater discharge, or the consumption of fish and shellfish that have been contaminated by pollutants. There is a strong correlation between swimmers in contaminated waters and higher rates of gastrointestinal illness compared to nonswimmers (Burton and Pitt, 2002). Beach closings and swimming advisories are often associated with elevated levels of bacteria indicative of human pathogens. For example, in 1996, approximately 83% of the beach closings and advisories were due to bacteria levels from stormwater discharge that exceeded beach water quality standards (NRDC,1997). In addition to bacteria causing a threat to the public, medical waste or sharp objects associated with gross solids can also be harmful. A study by HydroQual, Inc., Floatables Pilot Program in New York (2000) found that

1.7% (by number of items) of the gross solids captured were sensitive items, defined as syringes, crack vials, and diapers.

Gross solids are unsightly and may cause hazardous conditions for aquatic organisms and humans. When many people think of "pollution" they think of gross solids (or litter) because it is visual and may cause objectionable odors. Litter and debris can lower property values and be detrimental to tourism and recreation. Some gross solids can cause immediate human and animal harm, such as medical waste, sanitary items, glass, rusty metal, or sharp objects. Gross solids may potentially leach harmful pollutants into the stormwater, such as organic material leaching nutrients, toxic substances and other pollutants. Debris biodegrades requiring oxygen, therefore it reduces the amount of dissolved oxygen normally in the water. Although the impact on human recreation is recognized, stormwater solids have an even greater impact on the biology, geomorphology, and general water quality of the receiving waters.

The following sections summarize: 1) the impacts of solids on geomorphology by changing the substrate and channel forms; 2) the impacts on ecology and reasons that there is a general shift toward less sensitive species; and 3) how stormwater solids negatively impact water quality in receiving waterways by transporting harmful chemicals in runoff.

2.2 Geomorphic Impacts

Stormwater solids are known to cause various geomorphic impacts of concern because of the physical transport process through sedimentation and erosion. On one hand, nature maintains a dynamic balance among water yield, water velocity and depth, concentration and size of sediment moving with the water, width, depth, slope, hydraulic roughness and lateral movement of the stream channel (U.S.ASCE, 1995). On the other hand, urban development affects stream hydraulics and sediment input, transport, and deposition, and thereby alters the aquatic habitat and the resident community of aquatic organisms (Garie and McIntosh, 1986; Yoder and Rankin, 1997; Kennen, 1999; Paul and Meyer, 2001). There are many factors that may lead to channel instability, including channel and watershed slope, stream network configuration, base level, phase of urban development, distance from stream to urban land, riparian conditions, erodibility potential of the channel bed and banks, local sediment transport characteristics, proximity of geomorphic thresholds, and history of past disturbances (Bledsoe and Watson, 2001). In addition to the many watershed factors that lead to instability, stormwater runoff is a highly variable factor, both spatially and temporally. All of these factors combine to make the changes in geomorphology and stream stability difficult to predict.

As urban areas grow, so do impervious areas, resulting in increased peak flows and volumes of stormwater runoff occurring on a more frequent basis. Urban streams are therefore forced to accommodate this temporal rapid flow and volume. This increase in volume leads to channel instability, causing widening and scour in order to accommodate larger flows as a response to urban development (Hammer, 1972; Doll et al., 2002; Center for Watershed Protection, 2003). If the channel cannot physically widen, it will instead adjust by incising as a result of eroding the bed material. The introduction of large amounts of sediment with various particle sizes into the stream flow will also occur. The increase in flows results in erosion and instability and a tendency toward incising or widening depends most directly on the boundary material. A decrease in particle median diameter results in aggradations, a finer substrate, and reduces diverse habitat areas because of embeddedness. Urban land uses are positively correlated with fine sediment and embeddedness characteristics that have damaging effects on ecology.

Channel erosion is usually the predominant source of sediment in destabilized watersheds (Roesner and Bledsoe, 2003). Stream channel erosion and channel bank scour provide direct evidence of water quantity impacts caused by urban stormwater (U.S. EPA, 2003).

2.3 Biologic Impacts

The adjustment in solids in a fluvial system affected by urbanization is known to alter the quality and quantity of aquatic habitats. The ecosystem is affected by changes in habitat, oxygen availability, food sources, predator-prey relationships, competition, and behavioral patterns. Increase in sediment delivery can destroy aquatic habitat through deposition of fine sediments and destruction of important morphologic forms such as pool-riffle sequences (ASCE, 1992; Lenat et al., 1981; Lenat, 1984; Walters, 1995; Snodgrass et al., 1997; Simons and Senturk, 1991). It has been shown that biotic integrity begins to degrade even at relatively low levels of urbanization (Booth and Reinelt, 1993; Booth and Jackson, 1997; Maxted and Shaver, 1997; Wang and others 2000, 2001). Suspended solids cause behavioral effects, such as inability to see prey or feed normally; physiological effects, such as gill clogging; and reproductive and potentially lethal effects, such as burial and suffocation of eggs and benthic organisms, due to sediment deposition (U.S. EPA, 2003). The severity of the effect caused by suspended solids is a function of concentration, duration, particle size, associated toxins, and other stressors. It also depends on organism sensitivity, life stage, feeding habits and the frequency of exposure.

Sediment characteristics affect the distribution of fish species because of their tolerance for silty conditions. A change from gravel and cobble riffles to deposits of silt and sand results in both a reduction in the invertebrates most important as fish foods and a shift toward burrowing species as opposed to those inhabiting interstitial spaces (Walters, 1995). The biological communities become dominated by a smaller variety of species that are more tolerant to the degraded conditions (Allan, 1995; Herricks, 2001).

The abundance, richness, and diversity of stream invertebrates are closely related to the size and embeddedness of the stream bed sediments (Roesner, 1999). The percent embeddendness is the degree to which fine sediments such as sand, silt, and clay fill the spaces between rocks and a substrate. Evidence suggests that channel sediments may spend much more time in storage than in transport (Benda and Dune, 1997; Hadley and Schumm, 1961; Trimble, 1997; Walling, 1983) therefore settling to the bed. High percent embeddendness, where the rocks are surrounded and covered by fine material, results in blocking of spawning gravels. Sediments suffocate bottom dwelling organisms as well as smothering cobbles where fish lay eggs. Direct affects on invertebrates include abrasion, clogging of filtration mechanisms thereby interfering with ingestion and respiration and in extreme cases, smothering and burial, resulting in mortality (U.S. EPA, 2003).

Gross solids degrade the natural aquatic habitat. Small and large floatables can inhibit the growth of aquatic plants, decreasing spawning areas and habitats for aquatic organism. Settleable gross solids may smother productive sediments, bottom feeders, and benthic plants. In addition, wildlife living around the riparian area can ingest gross solids or become tangled in trash. Birds, mammals and sea turtles are vulnerable to entanglement and entrapment in plastic waste and litter. Some animals may ingest litter such as plastic bags which can lead to suffocation, intestinal blockage and death (Gaugler, 2003). It is estimated that some 100,000 marine mammals die every year from entanglement or ingestion of floatables and nearly a million seabirds are thought to die each year from entanglement or ingestion of floatable material (U.S. EPA, 2002).

2.4 Chemical Impacts

In addition to sediment being a major pollutant and causing harmful effects to stream geomorphology and physical habitats, many constituents use the sediment as a transport medium by adsorbing or absorbing to the solid. Harmful constituents, such as copper, zinc, cadmium, chromium, lead, and nickel (Gupta and Harrison, 1981; Horowitz, 1995), pesticides and phosphorus that are often detected in urban runoff have a strong affinity to particles. As particles decrease in size, they have a higher specific surface area and a net charge, giving them a greater capacity to sorb constituents on their surfaces such as metals and non-polar organics (Sansalone and Buchberger, 1997; Krein and Schorer, 2000; Stahre and Urbonas, 1990). Therefore, runoff containing higher concentration of particles from rain storms, snow storms, and periods of snow-melt are often associated with having a higher toxicity. The Nationwide Urban Runoff Program (NURP) (U.S. EPA, 1983) determined that heavy metals, especially copper, lead and zinc, are by far the most prevalent priority pollutants found in urban runoff (Harper, 1998). Metal concentrations are typically 100 times greater in urban runoff than in non-urban runoff (Welch, 1992). Trace elements have both natural sources from weathering of rock and soil, and anthropogenic sources such as combustion of fossil fuels, use of vehicles and road maintenance (Horowitz, 1995). Metals that have accumulated on settleable solids pose risks to bottom-feeding organisms. Metals can alter the reproductive rates and life spans of aquatic species. These contaminants can also accumulate in particulate feeder organisms and enter the food chain causing adverse effects on ecosystems in receiving water.

Nutrients in stormwater, such as phosphorus, can cause problems when in excess. In general, about 40-50% of nitrogen and phosphorus in runoff is in a dissolved form, while 50-60% exists in particulate form (Harper, 1998). Excess nutrients may result in algal blooms which favor survival of eutrophic aquatic species more tolerable to lower DO and high nutrient levels. Algal blooms reduce DO, may be toxic, and cause objectionable odors and tastes. Stormwater solids derived from street runoff is less readily biodegradable.

Other chemicals to consider in stormwater runoff that have direct negative effects include organics, oil, grease, xenobiotics and hydrocarbons. The COD/BOD_5 ratio for raw domestic sewage is about 2-3, while for street runoff it is 4-5 (Ashley et al., 2004), resulting in a larger impact on the dissolved oxygen in the water. Some chemicals can negatively affect reproduction, growth and development of aquatic species. Other chemicals may have highly toxic effects and cause chronic stress or kill aquatic species. Stormwater runoff from urban and some agricultural areas contain pharmaceuticals and other chemicals at concentrations that are a potential threat (Jones-Lee, 2005). There is a concern that pharmaceuticals and other emerging contaminants, although not necessarily toxic, may cause behavioral and biological effects having long term consequences on aquatic life and pose a threat to human health (Jones-Lee, 2005). Harmful pollutants can bioaccumulate in the tissues of fish and other species which may be very harmful (Harper, 1998) and persist in large concentrations in the food chain. Large particles and leaves also transport significant amounts of pollutants including polycylcic aromatic hydrocarbons (PAHs), metals, and nutrients.

Stormwater solids and their associated pollutants can infiltrate and negatively impact the quality of groundwater sources. For example constituents such as nutrients, pesticides, other organics, pathogens, metals, and salts and other dissolved minerals all may adversely affect groundwater quality (Pitt and others, 1994). Organics, metals and pesticides can cause environmental and ecological problems, as well as potentially harming human health. Suspended solids are of concern because of the potential for clogging the natural infiltration areas (Crites, 1985).

2.5 Impacts Summary

Stormwater solids can adversely impact the geomorphology, stream biology, and water quality. Some of the important environmental impacts are summarized in Table 2-1. For the purpose of classification in this document, gross solids are defined as solids greater than 5 mm, coarse solids are solids greater than 75 μm and less than 5mm, fine solids are solids greater than 2 μm and less than 75 μm, and dissolved solids are solids less then 2 μm (See Chapter 3.0).

Table 2-1. Possible Adverse Environmental Impacts of Stormwater Solids.

	Solids				
	Gross	**Coarse**	**Fine**	**Dissolved**	**Organic**
Size	Greater than 5 mm	Less than 5 mm, Greater than 75 microns	Less than 75 microns, Greater than 2 microns	Less than 2 microns	Estimated as loss upon ignition
Biologic					
Plants	May be abrasive and leach toxins inhibiting growth	May smother plants or be abrasive.	Increases turbidity and reducing sunlight penetration needed for photosynthesis	May transport toxins and nutrients that are harmful to plants. Dissolved solids impact color.	Increase turbidity, Reduced oxygen levels, especially at night needed for metabolism
Invertebrates	Smothers habitat zones and reduces hiding places from sedimentation. Less diverse biotic community (Paul and Meyer 2001)	May smother habitats for invertebrates and cover productive sediment. Can cause harm to bottom feeders through sedimentation and increased toxic material in bedload.	Infilling of fine sediment between spaces needed for habitat causing less divert invertebrate populations. Can affect feeding habits	Toxins and metals may bioaccumulate in invertebrates and persist in the food chain. Colloids and bacteria are potentially hazardous to the ecosystem.	Reduced dissolved oxygen may result in benthic community alteration to species more tolerant to low DO
Fish	Animals may ingest small pieces of trash such as plastics or become entangled in gross solids.	Smother fish eggs, reduce spawning success. Loss of hiding places for refugia by changing morphologic forms.	Visual impairement causes reduced feeding and production. Small particles also cause gill clogging.	Toxins and metals may bioaccumulate in invertebrates and persist in the food chain	Reduced oxygen causing oxygen sags and fish kills if introduced in high concentrations.
Other	Small solids may be eaten and cause harm to birds and other wildlife. Harm to humans and animals from sharp or entangling objects Can affect recreation and aesthetics.		Reduced habitat complexity may stress frogs, turtles, etc. This has an effect on water temperature, aquatic habitat, and water vegetation interaction.		
Geomorphic	Increase channel erosion and incision (Wolman 1967, Roberts 1989, Booth 1991), changes stream energy, channel forms, and mean particle size of the bed material. This in turn has an affect on the aquatic habitat, vegetation interaction, and water temperature				
Chemical	May leach harmful contaminants and organics may reduce dissolved oxygen in the system. Debris can result in large amounts of nutrients added to the water system. Gross and coarse solids can transport pathogens and bacteria		Transport harmful toxins, such as metal, toxic organics, and nutrients. Increased contaminated load and concentrations		Reduce the dissolved oxygen in the water which may increase the solubility of metals and phosphorus. Oil, greases, and hydrocarbons are very problematic in runoff. Organics used as a substrate for bacteria.

CHAPTER 3.0

CURRENT SOLIDS CLASSIFICATION

3.1 Introduction

Urban drainage transports dissolved, suspended, floating, washload, and settleable solids that may impair the ecology, geomorphology, and water quality of receiving water. This chapter reviews the existing solids classification for dissolved, colloidal, suspended and gross solids including classification by size and analytical protocol.

A clear method for consistent solids definitions, sampling, and measuring solids is deemed necessary for appropriate stormwater management. Therefore, an important goal of this chapter is to identify both commonly and less commonly utilized parameters and to detail their advantages and disadvantages as applied to stormwater. For example, TSS is a frequently reported parameter, but may not adequately represent stormwater pollutants because the correlation between TSS and specific pollutants is highly variable. Associated pollutants on the particulates may include heavy metals, phosphorus, nitrogen, pesticides, oxygen demanding substances (Woodward-Clyde, 1999) and other complex substances. The particle size and settling characteristics may be more important than the concentration of TSS because higher concentrations of pollutants have been found to be associated with finer particles (less than 100 μm). Fine particles (particle size less than 100 μm) have a tendency to agglomerate (Ashley et al., 2004). Sediment in stormwater varies greatly in size and specific gravity but may be divided into groups based on size and ability to settle or stay in suspension. Figure 3-1 is a size classification chart that depicts the current means by which solids may be classified. The following text describes the type of solids illustrated in Figure 3-1.

Total solids (TS) are defined as the mass of solid matter in the water, including both dissolved and suspended components. Total Solids consist of everything that is in water except for dissolved gases and the water itself. In stormwater, TS will typically include particles derived from rocks, biological material, chemical precipitates, pavement dust and particles, atmospheric dust, natural soils, vehicle rust particles, brake pad and tire dust and particles, trash, plant and leaf material (James, 2003). Many heavy metals and other trace elements are associated with sediments that comprise a significant portion of the total solids (Bent et al., 2001) although the correlation between suspended sediment and specific pollutants varies. Total solids may be further categorized into dissolved, suspended, settleable, fixed and volatile solids, as described in the following sections.

There is significant interest in the suspended versus dissolved components of solids, as the definition greatly affects the expected impact on habitat and potential options for treatment. Suspended solids include organic matter and inorganic matter. Sediments typically make up the majority of suspended solids in stormwater and may originate from soil erosion, including inorganic solids from weathering of rocks, or from locations where organic content has been reduced by decomposition. Organic solids tend to be less dense than inorganic solids because inorganic minerals are typically denser than organic carbon. In addition, fresh organic matter has

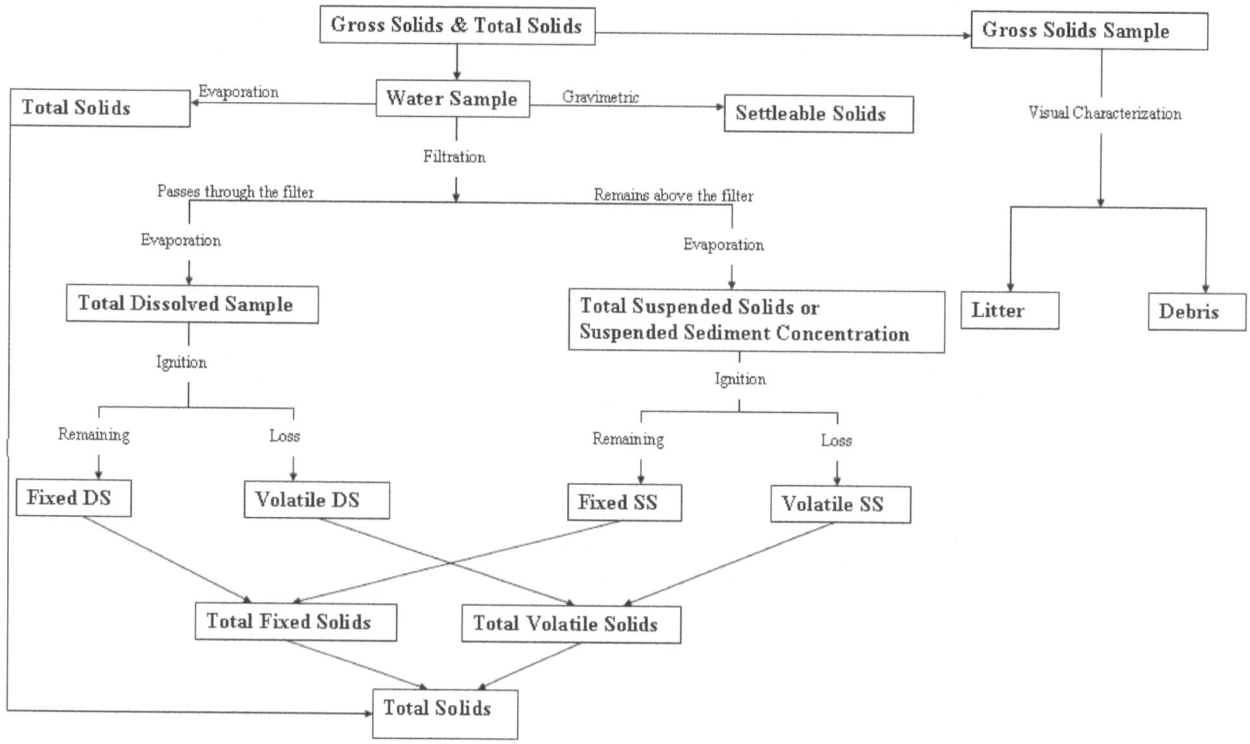

Figure 3-1 Current Solids Characterization.

high water content, causing its gross density very close to that of water (Chapra and Pelletier, 2004) and thus having a tendency to settle more slowly. Suspended solids are typically made up of fine material and solids with a high organic content.

It has been reported that total surface area per unit of mass increases with decreasing particle size (Sansalone and Tribouillard, 1999; Sansalone et al., 1998) therefore the finer the particles, the larger the surface area per unit mass for sorbing chemicals, thus the higher the associated pollutant load (Ashley et al., 2004). Because of this high specific surface area, fine material has the potential to transport the majority of chemicals such as nutrients, metals, and harmful organic compounds. For example, a study by Ellis and Revitt (1982) found that 70% of the metallic pollution is attached to particles finer than 100 μm which represent less than 15% of the total solids mass accumulated on streets. Methods for determining the amount of suspended solid-phase material in the water include *Total Suspended Solids (TSS)* and *Suspended Sediment Concentration (SSC)*. Although stormwater suspended solids values are highly variable temporally and spatially, a benchmark value of 100 mg/L has been selected by the EPA for TSS as the median concentration from the National Urban Runoff Program (Dodson, 1999). Thus, a value of 400 mg/L would be considered to be at the high end of typical stormwater runoff suspended solids (Dodson, 1999). It is important to note that even though SSC and TSS values are often used interchangeably, the analytical methods are very different and therefore the two terms should not be substituted for one another (Gray et al., 2000). *Settleable solids* are defined as the quantity of solids that will settle out of suspension in a water-solid mixture within a defined amount of time and can be determined by volumetric and gravimetric measurements. The ability for solids to settle out of suspension is a function of the particle density, shape, and size. In the United States, settleable solids are seldom reported in stormwater studies but have a huge impact on water quality, geomorphology, ecologic habitats, and treatability.

Volatile solids (VS) are generally considered to correspond to organic solids. However, it is important to note that VS measurements do not necessarily distinguish precisely between inorganic and organic matter because the loss of mass during ignition is not confined solely to organic matter, and may also include the decomposition or volatilization of some mineral salts (SM 2540A). However, the VS test is relatively simple to perform and is widely used as a surrogate for the direct measurement of organic material. Dissolved organic material, in particular, imparts color to the water and restricts light penetration. Because organic material is biodegradable, it also contributes to the biochemical oxygen demand (BOD) and therefore removes oxygen from the water by providing a readily available organic carbon source for aerobic respiring bacteria. Dissolved oxygen (DO) is the amount of oxygen dissolved in the water as a result of exposure to the atmosphere and photosynthesis and is a critical habitat variable for supporting aerobic aquatic life. Reduction of DO to less than 3-5 mg/l can cause a highly adverse impact to aquatic life forms that are sensitive to low DO levels.

Colloidal particles are solids that have a dimension roughly between 1 nm and 1 μm in at least one direction (Vignati et al., 2005). These particles are not dissolved, and therefore are solids in suspension; however, due to their colloidal nature, they do not settle. Colloids tend to scatter light. Organic or inorganic compounds may be present as, or associated with, colloidal solid phases in a continuous spectrum of size down to the nanometer scale (Grout et al., 1999). This material can include carbonate, bicarbonate, chloride, sulfate, phosphate, nitrate, calcium, magnesium, sodium, organic ions, and other ions. The fraction that is truly bio-available (in a form that can be taken up directly by microbes, plant or animals) is unknown (Davies, 1986; Welch et al., 1998). The biodegradable organic matter associated with particles larger than about 10^{-4} μm cannot be transported through the bacterial cell wall and therefore, from a microbial point of view, are considered as non-dissolved (Ashley et al., 2004). The particles have an outer layer of ions of the same charge, therefore they repel each other. These solids are usually not removed by mechanical devices and require the aid of chemical aggregation prior to settling, however they can be removed with very fine filters.

Studies show that although colloidal and fine particles may account for a small amount of total mass, they appear to be extremely important with respect to the transport and fate of metals, microorganisms, and organic compounds (Grout et al., 1999). Colloidal particles have a disproportionate influence on ecology and water quality because of their relatively high surface area, mobility, and bioavailability. Colloidal matter also constitutes a potential pathway of contaminants to filter feeders. Colloidal and fine particles are likely to stay in suspension contributing to turbidity, thus having a negative impact on sensitive rooted aquatic plants by causing reduced light availability.

Cohesive solids are a result of (Richardson et al., 2001)

- Ionic attraction between individual particles
- Type of mineral
- Particle spacing
- Salt concentration in the fluid

- Ionic valence
- Hydration and swelling properties of the constituent material

Cohesive soils are usually clay, (sizes less than 0.004-mm) or soils with a high clay content binding to silts and sand. The boundary between cohesive solids and non-cohesive solids is not always clearly defined although it can be noted that cohesion, the force by which particles are bound together, increases with decreasing particle size. Non-cohesive solids generally include sand, gravel, cobbles, or a combination of these.

Gross Solids (GS) include large particles such as litter, debris, and coarse sediment. Litter is defined as human derived solids such as plastic, aluminum, Styrofoam, wood, paper, glass, cigarette butts, and miscellaneous trash. These categories can be further divided based on the solids ability to float (floatable or non-floatable) and whether it is biodegradable or non-biodegradable. Debris includes organic material such as leaves, branches, seeds, twigs, and grass clippings. *Coarse sediment* is comprised of inorganic breakdown products from soils, pavement, and building materials. Gross solids are commonly transport as floatables (litter and debris) suspended within the water column or as bedload (coarse sediments). The minimum size limitation in the literature varies between 75 μm (Rushton et al., unpublished draft document, 2006) to 20 mm (Armitage and Rooseboom 2000a, 2000b) (Butler et al., 2002; Allison et al., 1998; Cornelius et al., 1994). A common definition of size that differentiates Dissolved, Fine, Coarse and Gross Solids is needed.

There is currently no standardized size range that constitutes gross solids. The California Department of Transportation defines gross solids as particles that are larger than 5 mm, while the Urban Water Resources Research Concil (UWRRC) Gross Solids Committee, (Rushton et al., 2006), suggests the lowest measured particle size for gross solids be 75 μm. Kayhanian and others (2005) defined litter as particles greater than 6.35 mm. A study by AMAFCA/ Albuquerque MS4 on floatable and gross pollutants in 2005 defined gross solids as particles greater than 1-3/4 in (45 mm) (Dodge, 2005).

The California Department of Transportation (2000) has done significant research on gross solids. They define litter as any man-made object that can be captured in a ¼-inch mesh screen (about 5-mm). This definition does not include materials of natural origin such as soils, gravel and vegetative debris. Examples of litter include cartons, wrappers, paper or plastic cups, cans, napkins, and cigarette butts. A study by Caltrans showed the following results:

- Between 75-87% of the gross pollutants by wet weight carried by stormwater are vegetative material (such as leaves and twigs)
- Plastics, styrofoam, and paper (including paper and cardboard) are the main components of stormwater litter materials monitored with very high numbers of cigarette butts also found
- Average annual litter loads monitored at drainage outfalls ranged from 3 kg (18 L) to 8 kg (58L) per acre of freeway surface, based on air-dried litter
- Average annual gross pollutant loads monitored at drainage outfalls ranged from 58 kg to 115 kg per acre of freeway surface by wet weight
- The moisture content of the litter was variable
- Material was deposited into the drainage system via both dry and wet weather processes

Caltrans recommends sampling for gross solids by placing a 5-mm mesh bag attached to the end of moderately sized outfalls (<610-mm/24-in). Sample collection can be done at the end of discharge pipes. Sampling in an open channel with solids that are transported across the entire cross section makes collecting a representative sample difficult. It should be noted that gross solids are continually degrading; therefore if captured materials are not removed and analyzed in a timely fashion, the measured fractions and sizes may be different than was contained in the fresh sample (Lenhart, and Lehman, 2006).

A summary of the sizes limitations used to define TDS, TSS and gross solids is shown in Table 3.1. The ranges used to define the various solids classification reinforce the need to develop a consistent definition.

Table 3-1. Summary of Size Limitations for Defining Stormwater Solids.

Size Classification	Total dissolved Solids	Total Suspended Solids	Gross Solids
lower limit	NA	0.45-2 μm	75 μm - 20 mm
upper limit	0.45-2 μm	75 μm - 20 mm	NA

CHAPTER 4.0

SAMPLE COLLECTION AND HANDLING

4.1 Sample Collection

It is very difficult to measure solids in stormwater because of the uncertainty in precipitation and weather patterns. Temporal and spatial variability in runoff result from a combination of factors, including volume and intensity of precipitation, rate of snowmelt, and features of the drainage basin such as area, slope, infiltration capacity, channel roughness, and storage characteristics (Bent et al., 2001). This section outlines sampling techniques and lists advantages and disadvantages of different techniques. Integrated and point samples are also identified in this section.

The largest problem with respect to sampling suspended solids in urban runoff may be obtaining a representative sample. Suspended solids vary with size, weight, time, and location. They also vary vertically with depth as well as horizontally. Larger particles, including gross solids (floating and suspended) and bed material, are often neglected in stormwater solids sampling because the equipment used cannot capture them. However, if the gross solids are not sampled, but remain in the capture facility, the organic fraction may decompose into smaller sizes and be monitored in the facility effluent perhaps introducing bias into the sampling results (Lenhart, 2006). Bed and gross pollutant loads may be twice the suspended solids load by mass and have significantly more volume because of their particle size (Hannah, 2005). Although bed and gross solids typically have a lower toxicity associated with them, they are important because they may cause premature failures of treatment devices that do not have a large enough storage capacity and may negatively impact stream habitats. Gross solids can also transport harmful chemicals into receiving waterways impacting water quality.

4.2 Sampling Technique

Suspended sediment is variable vertically in size, specific gravity and concentration from the water surface to the bed as well as laterally along the channel. The tendency for sediment to settle toward the bottom creates variability in the vertical direction, with higher sediment concentrations and specific gravity near the bottom and lower sediment concentrations near the surface. In addition, some solids tend to float therefore remaining at the surface of the water. This makes it difficult to get a representative sample of solids in the vertical direction. However, as a rule of thumb, fine material (silt and clay, less than 0.062 mm) is uniformly distributed from the surface to the bed (Bent et al., 2001; Edwards and Glysson, 1999).

Samples can be collected manually or with automated samplers (autosamplers). Manual samples can be taken with an open bottle, weighted bottle, and various sampling instruments. Automated suspended sediment samplers are limited with respect to the depth that the sample is taken due to the physical location of the perforated sampler tube relative to the bottom of the sampler which prevents sampling directly above the bed (Edwards and Glysson, 1999). The distance above the bed which is usually unmeasurable ranges from 3-1/2 inches to 6 inches (Chang, 1988). The size of the solid sampled is also limited by the nozzle size, neglecting larger particles such as litter, debris, and coarse sediment. Organic material, sediment, or ice may clog the nozzle.

Two advantages of manual collection are that it may be less costly than using an automated sampler (unless long term sampling at a specific location(s) is planned) and it is appropriate for all pollutants. A few disadvantages are that it is labor intensive, requiring personnel in the field to take the sample. Some locations are difficult to get to within the runoff period; therefore manual sampling would be difficult. In some cases, the conditions may be unsafe or impractical to sample manually. Human error in sampling is a greater possibility than for automated sampling. The procedure outlined in the NPDES Stormwater Sampling Guidance Document or similar documents by a comparable source should be followed when manually sampling.

Depth-Integrated samplers are designed to manually take samples as the container is lowered from the water surface to the bed and then raised back up to the surface at a uniform rate. Samples should be taken with the nozzle facing upstream. The samplers are equipped with a water inlet nozzle and an air outlet, allowing isokinetic sampling (the stream water approaching and entering the sampler intake does not change in velocity from the natural flow of the surface water) (Lane et al., 2003).

Non-isokenetic samplers include hand-held bottles, weighted-bottle samplers, the biochemical oxygen demand sampler and the volatile organic compound sampler (Lane et al., 2003). The hand-held bottle is the simplest form of sampling, where a bottle is dipped into the water to collect the sample. A sample scoop, which is a wide, flat plastic scoop, can be used to obtain the sample when the depth of flow is not sufficient to fill the sample container directly (Dodson, 1999). The BOD and VOC samplers are designed to collect samples without aerating the sample area.

Point-Integrated samplers are equipped with an electrically controlled remotely operated valve which opens and closes the sampler allowing the operator to isokinetically sample points on a given vertical (Lane et al., 2003). The sampler can be lowered to a depth prior to opening and taking a sample. These samplers can be used to obtain a time-integrated sample at a specific depth or can be moved vertically to obtain a depth-integrated sample. The samplers are usually equipped with changeable nozzles and bottle sizes dependent on the stream velocity, depth, and transit rate.

According to the Federal Inter-Agency Sedimentation Project Report 14 (FISP, 1963), point-integrating samplers are more versatile than the depth-integrated samplers. They can sample any point from the surface of the stream to a few inches off the bed as well as integrating over different depths. It should be noted that no specific sampler is best for all streams or all conditions.

Equal-discharge increment and *equal width increment* methods are used to sample in the lateral direction. If the water body is not completely mixed, a laterally integrated sample may be necessary. The equal discharge increment method divides the entire cross section into a minimum of four or a maximum of nine increments of equal discharge but unequal widths (Lane et al., 2003). Sampling is done at vertical locations within each of the equal-discharge-increments, usually at the location most closely representing the centroid in the increment or as a depth integrated sample. The concentration in all the verticals can be averaged to obtain the mean concentration over the entire cross section. The velocity distribution across the entire cross section must be known prior to sampling in order to select measurement points. The equal-discharge increment can save time and labor compared to the equal-width increment method because fewer verticals are required (Hubbell and Matejka, 1959).

The equal-width-increment method divides the stream into increments of equal horizontal widths but unequal discharges. It is recommended to use a minimum of 10 and a maximum of 20 intervals across the entire cross section (Edwards and Glysson, 1999). Using the same transit rate, depth–integrated samples are taken at each of the intervals. The sample volume will vary with each vertical sample because the depth and velocity will vary. The composite sample represents the mean cross sectional character of the suspended sediment.

A summary and evaluation of the quality of stormwater in Denver, CO, by the USGS (Bossong et al., 2006) indicates that there is a strong correlation between volume-weighted concentrations and time-weighted concentrations. This study evaluated 52 discrete samples collected during three storms and evaluated the correlation coefficients between the two methods for compositing samples. The correlation coefficients were always at least 0.88 showing strong correlation between the two weighting methods. In addition, streamflow measurements were investigated. The study evaluated 255 composite samples and in most cases, the overall percent difference between mean streamflow associated with composite samples classified as rising, falling, and event were less than 35%, and there were not large differences in mean streamflow for different portions of the storm hydrograph. It is noted that this study is site specific, and may not hold true in different locations and drainage basins.

Extensive sampling that involves integrated samples are very time consuming and expensive. A case study by the USGS in Denver, CO (Bossong et al., 2006) evaluated 14 paired-stormwater samples to assess if water obtained from a fixed sampling point in the stream cross section was representative of water in the entire cross section. The study compared samples collected using cross section and depth-integrating methods described above compared to samples collected by automated pump samplers collecting from a fixed sampling location. The results showed that the samples collected from a fixed point location using a pump sampler were similar to, but not always the same as, samples collected using methods that integrate vertically and horizontally in the channel cross section. But it supported the view that if the intake of the fixed location sampler is judiciously placed in a well mixed section, the extra time and costs of vertically and horizontally integrating samples may be avoidable.

Automatic samplers are a very common way to sample stormwater. Autosamplers can be either pumped samplers or passive samplers. The sampling can be preprogrammed and fully automated. Autosamplers collect water from a fixed point after a pre-determined stage threshold or other trigger is met. They can take grab samples or samples throughout a storm. The frequency of collection is on a timed basis for the anticipated length of the storm or a flow weighted basis such that equal volumes of water pass by the sampling point between each sample. The relative depth of the sampling point will be a function of stage. With human sampling, it may not be possible to be on site when the storm happens due to unpredictable weather patterns.

Pump samplers are usually equipped with a power supply, intake nozzle and pipe, pump, intake pipe, sample handling system and storage bottles, and a data logger (Chang, 1988). The intake should be placed at the approximate location of mean sediment concentration (which will vary during the runoff event). This is a source of error because the approximate mean sediment concentration is difficult to find and the relative depth of the sampling point will vary as a function of stage (Chang, 1988). Intake size is usually 3/8 in (9.5 mm) or 3/4 in (19 mm) (Edwards and Glysson, 1999), therefore neglecting larger particles (gross solids including litter, debris and coarse solids). The intake should be submerged all the time but should not be influenced by bedload material. The intake can be oriented five different ways listed below (Bent and others, 2001) and shown in Figure 4-1.

A. Normal upstream, directly into the flow
B. Normal and horizontal into the flow
C. Normal and vertical with the orifice up
D. Normal and vertical with the orifice down
E. Normal and pointing directly downstream

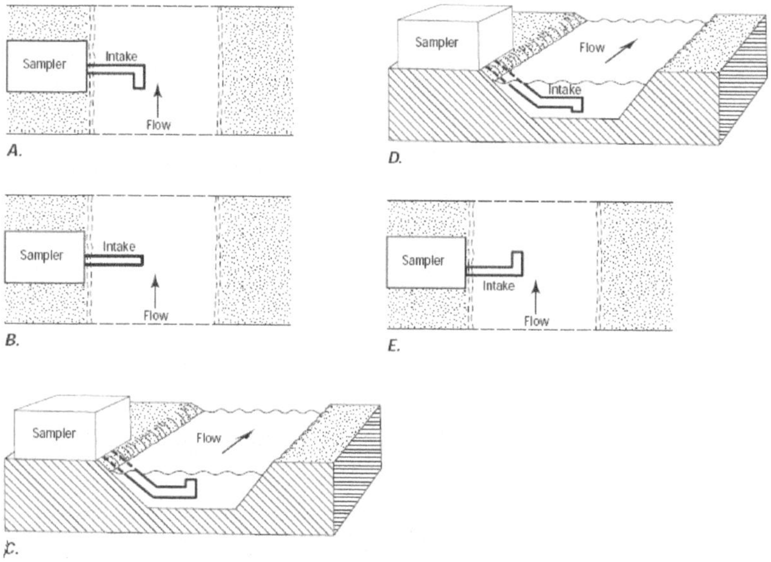

Figure 4-1. Pumped Autosampler Intake Orientation. From: Bent and others USGS Open-File Report 00-497 *A Synopsis of Technical Issues for Monitoring Sediment in Highway and Urban Runoff.* Figure 4 pp 14. http://ri.water.usgs.gov/fhwa/products/ofr00497.pdf

Orientation A, C, and D result in additional problems because they have shown high sampling errors and problems of trash plugging the intake. When the intake is pointed downstream (orientation E), a small eddy is formed directly behind the intake, which allows for a more representative sample of coarse load (Winterstein and Stefan, 1986). This orientation minimizes debris accumulation. In addition to intake orientation, the intake must have sufficient intake velocity to pick up and lift fine sand particles having settling velocities between 8 - 10 cm/sec range (Pisano, 1996).

Passive autosamplers are installed in the flow path and control the sampling rate by placement, orientation, and design of the water intake. Passive samplers generally include a sample intake, an inflow control assembly, a sampling container, and housing for the sample (Bent et al., 2001). There are a number of passive samplers explained in the 2001 USGS/FHWA study, A Synopsis of Technical Issues for Monitoring Sediment in Highway and Urban Runoff. The intake should be placed at a point that best represents the solid concentrations and characteristics in the runoff; however, this is very difficult to do. More than one intake can be used to try to get a more representative SSC and PSD. Most passive systems are not able to record flow when sampled and the intake may become clogged by debris. Some passive samplers are designed to collect bedload.

Autosamplers are convenient because they are already in use in monitoring programs and can collect discrete or flow-weighted composite samples throughout the duration of a storm. They allow for automatic and unattended collection of stormwater and require minimum labor. Autosamplers reduce the risk to personnel in unsafe conditions and they also reduce the risk of human error in sampling. This sampling technique is more costly than manual sampling because of the equipment required, and the maintenance and training on equipment that is necessary. Maintenance on the autosamplers includes checking equipment, changing sampling bottles, changing tubing and downloading data. Autosamplers are not typically capable of collecting an isokinetic sample, which may result in data that is not representative of the mean cross sectional concentrations and PSD, particularly when sand-size material is in transport (Edwards and Glysson, 1999). In addition, the size of the sample collected is often too small to be truly representative of solids in suspension in the water column. Automated sampling seems to be the biggest problem in acquiring a representative sample of stormwater. From the literature review, it is clear that the autosampler neglects bedload and gross solids which are major contributors to stormwater solids. Trash, litter, and debris are usually not sampled with the autosampler and are often times neglected because of this fact. A possible combination of suspended samplers and bedload samplers may produce a more representative sample, although it does not address the floating material.

Bedload sampling is commonly achieved using traps and pits designed to collect bedload material. Solids that are commonly found as bedload in urban runoff studies include the sediment retained on the road between storms, in drainage structures, in structural BMPs, or near drainage outfalls to receiving waters. Many of the bed material samplers designed for fluvial sampling are suitable for urban runoff settings (Bent et al., 2001). Streamside Systems, LLC, has developed innovative bedload monitoring collectors designed to capture fine bedload sediment. These samplers can be reviewed on the website: http://www.streamsidesystems.com/index4.htm. Sample efficiency varies with grain size. Traps are usually net-framed or pit traps anchored to the channel bottom that trap the solids moving along the bottom. Bedload trapping samplers are box or basket containers with the opening facing into the flow and a net that traps large solids. The net should be made of a durable material that has minimum impact on the flow (such as large mesh) and should be large enough to hold the volume of solids without compromising the solids trapped. Pit traps are designed to collect all solids that pass over them. This type of bedload sampler consists of containers set into the sediment with slot openings. As coarse solids pass over the trap the material falls through the slots and is trapped. A disadvantage to pit traps is they are usually quite difficult to remove. The opening of the trap should be large enough to allow solids to easily enter. It is also necessary to be aware that no bottom-material sampling equipment is appropriate for every objective and environmental setting (Radtke, 2005).

The errors associated with bedload sampling may be high because larger particles tend to move more irregularly under the influence of gravitational forces and are not well mixed in the water (Burton and Pitt, 2002). A long sampling time, or many short sampling times, is required to reduce bias. In addition, several bedload samplers should be used in close proximity because of the varied nature of bedload transport.

Bedload samplers and traps are often transportable, easy to operate, and cost effective. Bedload solids, as with suspended solids, are variable temporally and spatially. There are velocity (min and max) and depth limitations applying to both bedload samplers and suspended samplers. Bedload samplers do have particle size limitations depending on mesh size, size of sample mouth opening, and sampling location. Resuspension may occur from pits. Any sampler placed on the bedload will disrupt the natural flow and movement of solids.

Other sampling techniques include surface sampling utilizing vacuuming or washing solids off by water flushing. It should be noted that there is not a sampling technique that will sample all of the solids in the runoff other than to collect the entire flow.

4.3 Sample Types

There are two types of sample: grab and composite. Both types of sampling can be done with either manual collection or automatic collection. Grab sampling will not give a representation of event mean concentrations (EMCs) unless flow monitoring is also performed and the results are mathematically combined. Flow-weighted samples give better estimates of constituents than timed samples although of these methods composite samples give a better estimate of event mean concentrations than grab samples. Grab samples are simple to collect because no combination of samples is necessary, but grab samples are not usually comparable to flow weighted composite samples taken from the same storm. The collection must be in accordance with 40 CFR Part 136 for most permit compliance sampling.

A grab sample is an isolated, individual sample taken within a short period of time. It is a snapshot, showing the characteristics of the stormwater at a given time and discharge. The sample can be taken manually or by programmable autosamplers. The equal-width increment sampling technique or a single sample at the estimated centroid of the discharge is employed, so that it is representative of the majority of the discharge (Dodson, 1999). Grab samples are also commonly used for estimating the following constituents: pH, temperature, cyanide, total phenols, residual chlorine, oil and grease, total petroleum hydrocarbons (TPH), Escherichia coli, total coliform, fecal coliform, biochemical oxygen demand (BOD), chemical oxygen demand (COD), total organic carbon (TOC), and fecal streptococci.

Composite samples provide an estimate of the average concentration received over the entire storm. There are four methods to combine samples. Samples of *equal volume* are taken at *equal increments of time* and composited to make an average sample. Time-weighted composite samples are not acceptable, unless flow is monitored and the event mean concentration can be calculated from the data. The stormwater regulations require the collection of flow weighted composite samples, which can be accomplished by the following methods. First, samples can be taken at equal increments of time and can be composited proportional to the *volume of flow since the last sample*. Second, samples are taken at equal increments of time and composited based on the *flow rate at the time of sampling*. And finally, samples of equal volume are taken at *equal increments of flow volume* and combined.

Flow-weighted samples are a composite of samples taken throughout a storm and most commonly used when sampling stormwater. Samples are usually taken with an autosampler at predetermined time intervals during the duration of the storm. Flow-weighted samples must be taken for the first three hours or the entire event if it is less than three hours. The time interval must be at least 15 minutes apart or a minimum of three samples per hour. Ideally, the aliquots are obtained at constant time intervals of 20 min (Dodson, 1999). Water quality varies during different stages of the hydrograph which is why taking composite samples to collect a representative sample for storm events using flow weighted samples gives the best estimate of concentrations. To obtain the best representative sample, the sample must characterize solids concentration over the entire range of depth and a sufficient number of verticals must be taken to represent the solids variation horizontally.

4.4 Gross Solids Sampling

Currently, gross solids are often neglected in stormwater studies because of sampling practices. Most gross solids cannot be sampled by traditional automatic samplers because automatic samplers are limited to the collection of material smaller than the intake nozzle. Monitoring programs designed to determine the effectiveness of BMPs commonly report TSS without defining a maximum size of the solids.

Gross solids are usually sampled at a BMP that removes gross solids. These collection BMP devices including catch basin opening screen covers, curb or grate inlet catch basin inserts, hydrodynamic separators, end of pipe devices, linear radial gross solids removal devices, inclined screen gross solids removal devices, baffle boxes, litter and trash booms, and netting. A few of these devices that are used for gross solids removal are shown below in Figure 4-2 (Gordon and Zamist, 2005). These removal devices are not gross solids sampling devices, but a location where the gross solids are collected for removal and can be easily studied.

4.5 Handling

Important items to consider when sampling water-sediment mixtures are the type of bottle, how full the bottle needs to be, if preservations are required prior to sampling, and the maximum holding times for the sample. QA/QC procedures outlined in documents such as the USGS National Field Manual should be followed. Most samples should be collected in 1L bottles, leaving headspace (unless constituents being sampled require preservations added to the sampling bottle prior to sampling and no headspace). Sample bottles should not be overfilled. The samples should be immediately chilled to 4°C or below without freezing.

According to the procedures outlined by the NPDES, sample containers must be clean resistant-glass or plastic bottles and appropriate for stormwater collection. Preservation techniques guarantee that the sample remains representative of the original sample collected. Many constituents are unstable therefore preservation is necessary such as refrigeration, pH adjustment, and chemical fixation. Refrigeration at 4°C or below without freezing should be done at the beginning of collection in the field, continued through shipment and at the laboratory prior to analysis to minimize microbiological decomposition. It is best to collect and preserve samples right away. According to the Standard Methods for Examination of Water and Wastewater, it is preferable that the solid samples are analyzed within 24 hours but in no case should the samples be held for more than 7 days. Settleable solids have a maximum holding time of 48 hours (EPA 160.5). For other constituents refer to EPA 40 CFR section 136. Information such as rainfall intensity and duration, land use, soil type, seasonality, deicing practices, community events, street sweeping, etc. should be recorded.

CATCH BASIN INSERTS

Kristar FloGard Plus Catch

SCREEN COVER

Kristar FloGard Debris

END-OF-PIPE DEVICES

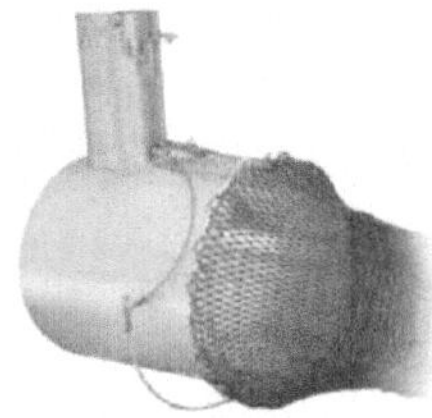

Kristar Net

HYDRODYNAMIC SEPARATORS

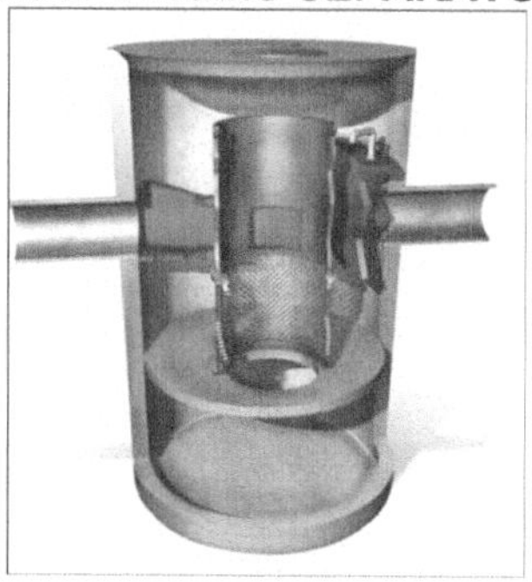

CDS Inline

NETTING

Fresh Creek Technologies

LITTER OR TRASH BOOMS

Kepner Plastics Sea Curtain

Figure 4-2 Gross Solids BMPs. From: http://www.plasticdebris.org/Trash_BMPs_for_Munis.pdf, 2005.

CHAPTER 5.0

SAMPLE ANALYSIS

5.1 Introduction

There is error and inconsistencies resulting from sample analysis in the laboratory. Some of the inconsistencies result from protocol and equipment. This section identifies current methods for analyzing suspended solids, gross solids, settleable solids and particle size distribution in stormwater runoff.

5.2 Total Suspended Solids vs. Suspended Sediment Concentration

Quantifying the suspended solid-phase material in water can be done using suspended-sediment concentration (SSC) and/or total suspended solids (TSS). In general, SSC values are calculated by measuring the dry weight of *all* the sediment in a known volume of water-sediment mixture. TSS data is produced by measuring the dry weight of sediment in a *subsample* taken from a known volume of water-sediment mixture. SSC and TSS values are often used interchangeably, but according to a study by the USGS (Gray et al., 2000), the values are not comparable and should not be substituted as each other. Also, there is no simple, straightforward way to adjust TSS data to estimate SSC (Glysson et al., 2000). In the study by the United States Geological Survey, the TSS method of analysis produces data that are negatively biased by 25% to 34% with respect to the SSC method (USGS, 2000), that is, TSS tends to underrepresent the actual mass of solids in the water column.

TSS analysis (EPA 160.2) requires withdrawal of an aliquot of the original sample, and this leads to inconsistencies in the methods used for TSS sample preparation. The subsample is filtered through a glass fiber filter (2.0μm or smaller, commonly 0.45 μm) and the residue retained on the filter is dried to constant weight at 103-105°C. Any sub-sampling usually increases the variance or creates a bias in the concentration and size distribution of solid-phase material. The error may come from the location from which the subsample is extracted (upper regions of samples have lower concentrations of sand while lower parts tend to be sand enriched) and also in withdrawal using a pipette (particles over 3mm may clog the tip of the pipette while small but heavy sand particles may precipitate out of the pipette as it is withdrawn from the main sample). Subsamples obtained by pouring are also unlikely to provide a representative sample, specifically in sand-size particles.

SSC values are calculated by measuring the dry weight of all the sediment in a known volume of water-sediment mixture. There is no subsampling, therefore SSC measures the suspended sediment in the entire sample. SSC can be determined using the method of evaporation, filtration, or wet-sieving filtration. All methods involve retaining, drying at 103-105°C and weighing the sediment in a known mass of water-sediment mixture. SSC reduces the error from sub-sampling because the mass of sediment and net weight is for the entire sample.

As noted above, tests performed by the USGS reveal that SSC values tend to exceed their corresponding paired TSS. Thus, TSS data used to calculate suspended-sediment loads in highways and urban runoff may be fundamentally unreliable (Bent et al., 2001) and therefore SSC may be more inclusive. However, the use of the SSC analytical procedure in stormwater management is not appropriate if the solids of concern are the lighter clays and organic material that have historically been the target solid for removal from stormwater. The difference in sediment concentrations measured is due to the analytical procedure of measuring the entire mass of sediment in the sample; therefore including solids that tend to settle (such as sand and small pebbles) and do not become completely mixed. Thus using SSC to calculate the removal efficiency of treatment devices leads to over estimates of percent removal of the historically important type of solids. However, SSC and TSS data may be comparable when the percentage or amount of sand-sized material in the sample is less than about 25% (Gray et al., 2000). In terms of consistency, the SSC method produces relatively reliable results regardless of the amount of sand-size material. The method for determining TSS, on the other hand, was originally designed for the analysis of wastewater samples and has been shown to be fundamentally unreliable for the analysis of the total solids in suspension in natural-water samples (Gray et al., 2000; Bent et al., 2001). SSC may be more inclusive of suspended solids captured using an autosampler, but it still does not address the larger solids missed by the autosampler (Rushton and England, 2006). A few other disadvantages of using SSC instead of TSS are that it is not widely performed by most laboratories and that the entire sample volume must be analyzed, therefore requiring more time and money. In addition, larger sample volumes would be required if SSC was used to analyze stormwater solids if other tests are to be carried out (such as analysis of metal or nutrient concentrations).

Indirect methods for measuring suspended sediment, such as turbidity, can often be less expensive to collect or analyze, however their relationship to SSC has not been defined. Therefore, surrogate techniques such as turbidity were not addressed in this research.

5.2.1 Filter Sizes

Separating the solid phase material from the liquid, or "dissolved", phase requires a filtering process. TSS are that portion of the solids retained by the filter while total dissolved solids are defined as the solids that pass through the filter. The filter nominal pore size, porosity, area, and thickness become a significant factor in which material is retained above the filter and which material passes through. The standard methods (APHA, AWWA, WEF, 1998) define nominal pore sizes for the filter as 2.0 μm (or smaller) and lists several appropriate filters: Whatman grade 934AH; Gelman type A/E; Millipore type AP40; E-D Scientific Specialties grade 161; Environmental Express Pro Weigh; or other products that give demonstrably equivalent results. It is noted that because of the physical nature of glass fiber filters, the absolute pore size cannot be precisely controlled or measured. Commercial filter diameters range from 2.2 to 12.5 cm.

The method for TSS subsampling under this protocol requires stirring the sample at a speed great enough to shear larger particles, if practical, to obtain a more uniform (preferably homogeneous) particle size (SM 2540 D). While stirring, a measured volume is pipetted from the approximate mid-depth and midway between the wall and vortex. The method dictates that samples should be kept homogeneous during transfer, by mixing small samples with a magnetic stirrer, and using wide-bore pipettes. Although, it is warned that "sampling, subsampling, and pipetting two-phases of three-phase samples may introduce serious error," subsampling with a pipette is still the recommended approach.

A few of the recommended filters in the above testing methods are summarized below.

- ♦ Whatman grade 934-AH <u>1.5 µm</u> - smooth surface, high retention borosilicate glass microfiber filter that withstands temperatures over 500°C. Specified in standard methods for determining TSS in water.
- ♦ GF/C <u>1.2 µm</u> - combines fine particle retention with good filtration rate. Used in many parts of the world for collection of suspended solids in potable water and natural and industrial waste. Widely used in biochemistry for cell harvesting, liquid scintillation counting and binding assays.
- ♦ Gleman Type A/E - Fine porosity, fast filtration rate, with a <u>1.0µm</u> size particle retention.

5.3 Laboratory Analysis

Analysis techniques follow standards such as the Standard Method for Examining Water and Wastewater, ASTM, and U.S. EPA and are summarized in Table 5-1.

Table 5-1. Solids Testing Methods.

	Test	Summary	Test Volume	Detection
Filterable (TDS)	EPA 160.1	Gravimetric: A sample is filtered through a standard filter and evaporated and dried at 180 °C	100 mL or larger	Total should be limited to 200 mg to prevent crusting
	SM 2540 C		volume to yield a residue between 2.5-200 mg	2.5-200 mg
Nonfilterable (TSS)	EPA 160.2	Gravimetric: A sample is filtered through a standard filter and the residue retained on the filtered is dried at 103-105°C	100 mL or larger	must capture at least 1.0 mg of residue.
	SM 2540 D		volume to yield a residue between 2.5-200 mg	No more than 200 mg
Suspended Sediment Concentration (SSC)	ASTM D3977-97 Method A	The sample is evaporated until dry	0.2-20L, entire sample	
	ASTM D3977-97 Method B	The entire sample is filtered. The filter and retained solids is dried	entire sample	<10,000 mg/L of sand and <200 mg/L of clay
	ASTM D3977-97 Method C	The entire sample is sieved with 62 or 63 um openings. The retained is weighed	entire sample	
Total	EPA 160.3	Gravimetric: An aliquot fo the sample is evaporated to dryness at 103-105°C	Volume to yield a residue of at least 25 mg	25 mg
	SM 2540 B		volume to yield a residue between 2.5-200 mg	No more than 200 mg
Volatile	EPA 160.4	The residue of total, filterable, or non-filterable residue is ignited at 550°C. The weight lost is volatile residue	volume used for TS, TSS, TDS	
	SM 208E		volume used for TS, TSS, TDS	
Settlable	EPA 160.5	Volumetric: Settle solids in an Imhoff cone and estimate the volume of settled solids in a predetermined time. Gravametric: Determine TSS, let settle and determine the TSS (nonsettlable solids)	1-L	0.2 mL/L/hr
	SM 2540 F: Volumetric		1-L	Depends on sample composition - usually 0.1 to 1.0 mL/L
	SM 2540 F: Gravametric		not less than 1-L, siphon 250-mL from the center of container to determine nonsettleable solids	Depends on sample composition - usually 0.1 to 1.0 mL/L

Samples are split primarily using a polypropylene/polyethylene or fluorocarbon polymer churn splitter, which also serves as a composite device, or a cone splitter, which requires a separate compositing vessel. Studies by the Office of Water Quality (Rickert, 1997) show that either device can be used for suspended-sediment concentration (SSC) less than 1,000 mg/L and the cone splitter should be used for samples with SSC concentrations less than 10,000 mg/L. The results were observed after running several laboratory tests using a predetermined amount of monomineralic silica sands in distilled water. In addition to larger concentrations, the cone splitter is more suitable for smaller particles. Cone splitters are much more effective than churn splitters when suspended solids and particle size analyses are critical (Pitt et al., 2002). Table 5-2 below summarizes the advantages and limitations of the churn and cone splitters.

Table 5-2. Churn and Cone Splitter.

Splitter	Advantages	Limitations
Fluoropolymer Churn Splitter	Can be used for inorganic and non-volatile organics Easy to clean No modification of design necessary	Splitting accuracy is unacceptable for particle sizes > 250 um Accuracy is unacceptable for SSC > 1,000 mg/L Sample volumes < 4L or > 13 L cannot be split for whole-water subsamples
Plastic Churn Splitter	Easy to clean Simple to operate	Must be used to composite samples for organic compounds Splitting accuracy is unacceptable for particle sizes > 250 um and SSC > 1,000 mg/L Sample volumes < 4L or > 13 L cannot be split for whole-water subsamples Requires a modified spigot and construction of a funnel assembly
Fluorocarbon-polymer Cone Splitter	Acceptable with SSC 0-10,000 mg/L Acceptable with very fine clay and silt to sand-size particles samples as small as 250 mL can be split into 10 equal subsamples Samples > 13L can be processed for both inorganic and nonvolatile organics	Accuracy of the volume equivalents must be verified before using a new or modified cone splitter Splitter is awkward to operate and clean Sample is vulnerable to contamination from atmospheric sources or from improper operation.

(USGS Field Manual Section 2.2 Table 2-6 pp. 58)

5.4 Particle Size Distribution

Particle Size Distribution (PSD) is useful for studying the chemistry, transport, and fate of sediment in urban runoff, BMPs and receiving waters (Kobriger and Geinopolos, 1984). As particle size decreases, the relative specific surface area increases, giving the particles a greater capacity to sorb constituents that accumulate at surfaces such as metals and non-polar organics (Sansalone and Buchberger, 1997; Krein and Schorer, 2000). Therefore, runoff containing higher concentrations of fine particles are often associated with higher toxicity. While suspended solids are commonly used as a surrogate for water quality, the quality and quantity of the solids should be addressed by determining the particle size distribution and analyzing the chemistry. For example, 80% TSS load removal of large particles would not have the same water quality outcome as 80% removal of a fine material. Therefore, PSD is an important design consideration for water quality improvement due to its correlation with sediment pollutant availability (Wong et al., 2000). One medium-size sand grain has the equivalent mass of about 1,000 medium-size silt grains given equal densities (Bent et al., 2001) therefore percent removal by weight may not properly address the issue of particle surface area which can be critical to water quality and ecology.

Particle size is a measurable property that may be important to study in stormwater solids in order to better understand the effects, sources and treatability of stormwater solids. This proves to be difficult because stormwater particles are in a dynamic state transforming through aggregation, flocculation, biodegradation, and disaggregation (Ashley et al., 2004). The characteristics of particles and associated pollutants throughout a range of PSDs offer a new approach for the characterization of stormwater pollutants (James, 2003). There are several methods for obtaining a particle size distribution for solids: sieve analysis, hydrometer, visual accumulation tube analysis, pipette analysis, light diffraction, light obscuration, electrical resistance counters and sedimentographs. Some researchers use a sieve analysis to separate the larger particles and then use another technique to analyze the finer particles. It should be noted that no method is ideal for all applications. These tests are summarized below (Pechacek, 1993).

- Sieve Analysis – generally used for particles greater than a No. 200 (75 μm) sieve. A solids sample is passed through a series of screens of progressively smaller openings, and the material caught on each sieve is collected and quantified. This can be performed on either wet or dry sieves.
- Hydrometer – used for particle sizes 1-75 μm (ASTM D422, 1992). This test uses Stokes Law to compute settling velocity. A hydrometer bulb measures specific gravity of the water-solids mixture at different time intervals and is calibrated to a soil with a specific gravity equal to 2.65. This test is slow but is applicable to describe transport properties and sedimentation.
- Visual accumulation (VA) tube analysis – used for particles diameters of 62 μm to 2 mm (usually sands). The solids sample is added at the top of a settling tube and allowed to stratify according to the settling velocities. There is a continuous trace of the deposited sediment at the bottom of the VA tube.
- Pipette analysis – used to determine particle size gradation for sizes 62 μm and smaller. The principle of this analysis is to determine the suspension concentration at predetermined depths as a function of settling time. An assumption of specific gravity equal to 2.65 is made for the solids to determine settling velocities.
- Light scattering counters (laser diffraction) – used for particle sizes from 0.1 to 50 μm. A laser illuminates a sensing zone as a sample flows through a sensor. Photodetectors measure the light scattered over a fixed angle. The principle is that smaller particles scatter the light waves through a wider angle than larger particles.

- Light Obscuration Counters – used for particle sizes 1 to 500 µm, depending on instrument capability. A light beam with a known intensity is detected by a photodetector. As particles pass through the zone, the beam is blocked and there is a decrease in the beam light.
- Coulter counters (Electrical resistance counters) – This method measures the electric current between two electrodes located on either of an aperture tube. The suspended solids cause a decrease in the current between the electrodes which produces an electrical impulse proportional to the volume of the particles in the aperture. By counting and sizing the pulses, it is possible to determine the size distribution of the suspended particles.
- Sedimentograph (x-ray sedimentation) – used for particle sizes 0.1 to 300 µm. The Sedimentograph is an x-ray based sediment size analysis instrument for use in the silt-clay size range. By measuring the gravity-induced settling velocities, based on stokes law, and mass fraction determined by a relatively low absorption of low-energy x-rays.
- Microscopy – Microscopy examinations can be done on particle sizes ranging from 0.5-5000 micron particles with a light microscope and 0.01-10 µm with a scanning electron or transmission electron (SEM/TEM) microscope.

Separation between sand and silt can be done by wet-sieving using distilled water and a 250-mesh (0.063 mm) sieve. This is the separation between coarse material (greater than 0.063mm) which are non cohesive and fine grained material which are cohesive. Large surface to volume ratio can mediate partitioning and transport of metal elements while serving as reservoirs for many reactive constituents (Sansalone et al., 1998) and specific surface area increases with decreasing size. Silt and clay sized particles (particle size less than 0.062 mm) are primarily responsible for transport of contaminants, silting of fish spawning beds, and disturbance of habitats for benthic organisms. These particles are more difficult to treat because of their small size and lower settling velocities.

Standard methods used in soil mechanics to determine particle size distribution include pre-drying and shaking through a series of sieves. This method may alter the original size and shape of the particle therefore wet sieving may be more appropriate. The visual accumulation tube method is used for sand sized particles by allowing particles to fall from a common source. PSD determined by settling columns is commonly used but may be problematic because many larger particles settle before the test actually begins (James, 2003). This problem can be avoided if samples are taken immediately upon filling settlement tubes. The pipette and settling methods rely on Stokes Law which assumes that the particles are spherical and have a specific gravity of 2.65 (Guy, 1969). The particles in stormwater are not all spherical and have a large range of specific gravities.

The classification accepted by the Sediment Terminology of the American Geophysical Union can be seen in Table 5-3.

Table 5-3. Particle Size Distribution (PSD) Classification (American Geophysical Union Sediment Classification System) (Lane, 1947).

Size (mm)	Class
64-32	Very coarse gravel
32-16	Coarse gravel
16-8	Medium gravel
8-4	Fine gravel
4-2	Very fine gravel
2-1	very coarse sand
1-0.5	coarse sand
0.5-0.25	medium sand
0.25-.0125	fine sand
0.125-0.062	very fine sand
0.062-0.031	coarse silt
0.031-0.016	medium silt
0.016-0.008	fine silt
0.008-0.004	very fine silt
0.004-0.002	coarse clay
0.002-0.001	medium clay
0.001-0.0005	fine clay
0.0005-0.0002	very fine clay

PSD in stormwater can change rapidly. Particles naturally grow and aggregate causing significant changes in PSD over time periods of hours. Solids undergo significant physical, chemical and biological processes as they are transported in runoff (Ashley et al., 2004). Because of this, it is recommended to analyze particle size distribution within six hours of sampling (including holding time in an autosample) which may not be practical in many stormwater studies. Online particle size distribution measurements in stormwater runoff can be used along with autosamplers to obtain data on particle size of solids in stormwater (Kayhanian, 2006). Research continues to develop in-situ water quality measuring devices providing alternative methods to obtain PSD information in urban runoff.

5.5 Particle Settling Velocity

Particle settling velocity is dependent on particle size, shape, specific gravity, and fluid viscosity. The kinds of solids in stormwater are affected by the type of soil in the watershed. Settling velocities of solids have a large range because they are dependent on the kind of particles in the runoff. In addition, stormwater particles tend to coagulate/flocculate over time. This naturally occurring flocculation and increase in particle size can influence stormwater solids settling velocities. Common stormwater removal practices rely on settling or filtration. Particle settling velocity is needed to effectively design BMPs to treat stormwater solids. See Ashley et al. (2004) Scientific and Technical Report for settling velocity devices.

The settling velocities of discrete particles are shown in Figure 5-1, based on Stokes and Newton's settling relationships. This figure also illustrates the effects of different specific gravities on the settling rates (Morquecho, 2005).

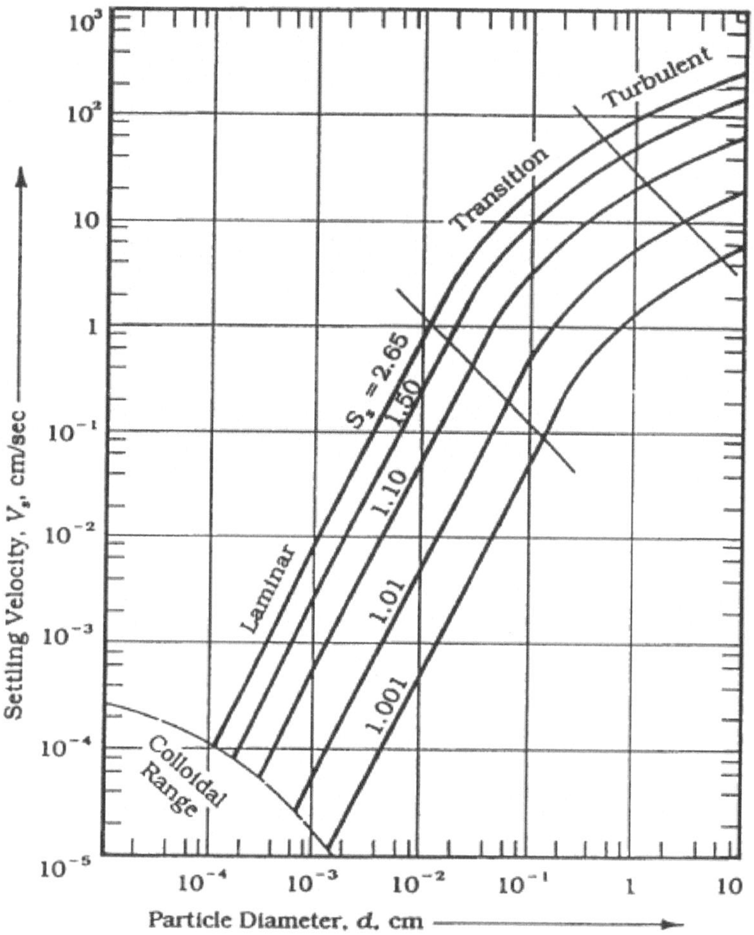

Figure 5-1. Settling of Spheres in Water at 10°C. From Unit Operations and Processes in Environmental Engineering, 2nd edition by Reynolds/Richards. 1996. Reprinted with permission of Nelson, a division of Thomson Learning: www.thomsonrights.com. Fax 800-730-2215.

In the above figure, laminar flow is flow that is not turbulent. It is a smooth, streamline type of viscous fluid motion characteristic of flow in parallel lines. This would represent water that is not in motion, such as in a stormwater detention pond. Turbulent flow is fluid flow in which the fluid undergoes irregular fluctuations, or mixing. A small amount of agitation may cause solids to remain in suspension or become resuspended. The flow would be turbulent during stormwater runoff.

5.6 Gross Solids Analysis

There is no common technique for analyzing gross solids and they are commonly neglected in stormwater solids analysis because of the current lack of accurate sampling techniques. Although neglected, gross solids may be a large contributor to stormwater pollutants.

Caltrans recommends a specific procedure to analyze gross solids (Caltrans, 2003). The total wet weight and volume of gross solids should be measured in the field. Gross solids can be weighed using an electronic scale while volume measurements can be done by displacing a known volume of water. Once the gross solids arrive in the lab, they should be separated into debris (vegetative matter) and litter. Separation may require visual examination and separation of natural and manmade material. The weight and volumes of each component should be recorded. The litter is allowed to air dry for 24 hours and again the weight and volume are measured.

The litter management pilot study (LMPS) by Caltrans characterized gross solids by weight, volume, and number of items. Air-dried weight was obtained using a digital scale; volume was estimated by placing the litter samples into graduated containers; and the number of items was determined by manual count. In this study, wet weights and volumes of the gross solids collected in the ¼-inch mesh bags was recorded upon arrival at the laboratory. The litter was separated from the vegetation and placed on drying racks where it was allowed to dry for 24 hours. The litter was then classified in the following categories. Each type of litter was further divided into prior usage categories—food related, smoking related, and other (Lippner et al., 2000):

- Cardboard/chipboard
- Moldable plastic
- Paper
- Plastic film
- Glass
- Styrofoam
- Metal
- Wood debris
- Cloth
- Cigarette butts
- "Other"

Similarly, in a study by HydroQual, Inc. in New York City, floatable litter characteristics were investigated. The categories of litter were similar to the LMPS study including plastics, metals, paper, wood, polystyrene, clothes/fabric, sensitive items, miscellaneous, and glass. In a floatable and gross pollutant study in Albuquerque (Dodge, 2005) gross solids were characterized by visual examination and the constituents of the debris leachate were also investigated. The categories used in this study were natural materials (including organics, leaf litter, tumble weeds, and grasses) and manmade materials (paper, fabric, cigarettes, lumber, cans, glass and expanded foam, bottles, bags, and plastic sheets). The volume was estimated using a graduated bucket and percent by volume for each category.

A leachate analysis was conducted on the gross solids collected in the AMAFCA/Albuquerque study. Leachate from the solids was used to determine the amount of water quality constituents of concern that would leach or wash off of the gross pollutants. Based on this study, it appears that a significant amount of water quality pollutants easily wash off of debris. The following procedure was used.

- One cubic yard of gross solids was collected from separate locations
- Each cubic yard was immersed in distilled water for 15 minutes
- The leachate was sampled and delivered to a laboratory for water quality analysis using standard procedures.

Their results supported the conclusion that gross solids can markedly degrade the water quality by increasing metal concentrations. Specifically total copper, total lead, and total zinc of the leachate solutions showed increased concentrations when compared to the stormwater average. The leachate also showed presence of Petroleum Hydrocarbons and Solvents, (ethylbenzene, toluene, total xylenes, 1,2,4-trimethylbenzene, 1,3,5-trimethylbenzene, chrysene, fluoranthene, and pyrene).

In addition to characterizing gross solids into several categories, Orange County Stormwater Program in California determines the water content of the gross solids. This is done by collecting a small subsample, placing it into an aluminum tray and recording the wet weight. The sample is then allowed to dry in an oven at 60°C for 24 hours. It is then reweighed to determine the dry weight. The difference in the wet and dry weights is the amount of water associated with the solids.

Currently, there is not a standardized way to sample and handle gross solids. There are large variations in reported numbers due to inconsistencies in how debris is collected for measurement (wet or dry storage, compacted or loose, volume or mass-based). There is not a common language for gross solids, which affects the way they are analyzed. These inconsistencies also make it difficult to compare results from location to location.

A principal source of error in testing gross solids in stormwater is the failure to obtain a representative sample in the field and in the laboratory. In the field, the traditional techniques of sampling using an automated sampler neglects bedload and fails to sample larger particles, proving to be a large contributor to the failure to obtain representative samples. In the laboratory, sources of error may come from keeping samples with multi-phases homogenous and subsampling from them. Part of the sample may adhere to the sample container or magnetic particles can stick to magnetic stirrers. When measuring solids using an evaporation technique, the sample may dry with the formation of crust that prevents water from evaporating or, in the case of drying gross solids under ambient conditions, the samples may take on water from humidity.

CHAPTER 6.0

LITERATURE REVIEW SYNTHESIS

6.1 Discussion

Stormwater greatly impacts the environment by altering the natural geomorphology, changing aquatic habitats, and carrying chemicals into waterways. In order to minimize these impacts, it is important to have consistent definitions of solids such as suspended, settleable, and gross solids. Dependable sampling and analysis protocols are also needed in order to assist in stormwater management. The following outline summarizes some of the most immediate concerns with respect to stormwater solids.

- There is currently not a single internationally recognized standard for classification of solids (dissolved, suspended, gross, settleable, floatable). This makes it problematic to compare results from different research and to implement stormwater management techniques. In particular, data regarding gross solids sampling, handling and analysis is limited and often inconsistent. The size limitations in the literature vary between 75 μm to 20 mm. A common definition and method for sampling and analyzing of these solids is needed.
- Field sampling equipment and sample collection protocols are inconsistent from study to study (and often site to site) with respect to how bedload and larger diameter solids are dealt with, and therefore do not give an accurate representation of these solids in the water.
- The location of the sample collection point adds error because the type and concentration of solids are variable vertically as well as horizontally. Nozzle location and orientation are important factors to consider when sampling. In addition, the sample intake velocities affect the solids that are collected. Optimally, samples should be depth integrated to account for the variation from the surface to the stream bed. For best results, samples need to be taken across the reach (Equal Width or Equal Discharge Integrated) to achieve a laterally integrated sample. If only one location can be sampled, samples should be collected at the vertical and horizontal point that is most representative of the centroid of the solid mass in the discharge.
- The irregular intensities of rainfall make it difficult to predict runoff rate, pollutant transport, sediment deposition and re-suspension, channel scour, etc. Pollutant sources in the area between storms such as landscape practices, spills, construction activities, and vehicle use dramatically alter the solids load in terms of mass, timing and composition. It is not surprising that pollutant characteristics can be very different from location to location as well as from storm event to event. Therefore, it is difficult to target one size of solid as the most problematic. Site-specific investigation should be performed and a specialized monitoring plan developed to meet the specific stormwater management goals.

- Testing Total Solids (TS) as the only constituent does not adequately represent the stormwater pollutants because they are highly variable in time, space, and particle size as well as the variability of pollutant concentrations associated with the particles. Removing large particles which may be the majority of mass is not the same as removing fine particles which carry a high percentage of harmful pollutants. Fine and dissolved solids have particularly significant negative impacts and should be considered when monitoring stormwater runoff.
- Particle size distribution is important to analyze because smaller particles are often associated with high toxicity and pathogenic potential. The characterization of particles and associated pollutants throughout a range of PSD offers a new approach for the characterization of stormwater pollutants. The size distribution and quality characteristics of solids in stormwater are site specific and difficult to quantify because they are variable temporally and spatially. There is no one solution for characterizing and remediating all stormwater pollution.
- Suspended Sediment Concentration (SSC) provides a more consistent and reproducible result for the solids in water than Total Suspended Solids (TSS) because of the error associated with subsampling. TSS usually under-predicts the solids actually in the sample. But there is a problem with changing the standard test to SSC because a relatively small proportion of larger particles can comprise the majority of mass and volume of the total solids in the sample. Using SSC in place of TSS analysis may allow for percent removal efficiencies to be met by capturing only larger particles, which will not result in the same water quality impacts as removing the smaller particles that are associated with transporting harmful constituents to the waterways.
- The filter size used to distinguish "TSS" from "Total Dissolved Solids (TDS)" is not consistent. An inconsistent filter size is a problem because the use of a standard 2 micron filter will produce different results than using a 0.45 micron filter, but any size smaller than 2 µm is acceptable according to the APHA Standard methods for the Examination of Water and Wastewater (SM 2540) test protocol.

Stormwater solids present a complicated challenge that requires site specific considerations. It is undisputed that stormwater is an environmental concern for receiving waters by government agencies around the world. The solids in stormwater carry high concentrations of harmful chemicals damaging to the ecosystem and the overall quality of water. Characterizing problematic solids and thereby developing reliable removal practices is necessary for effective stormwater management.

CHAPTER 7.0

Proposed Stormwater Solids Classification

7.1 Solids Definition

Stormwater-borne solids include many different types and sizes of solid material ranging from fine suspended sediment to bed-load; trash, coarse (gross) solids, and floating debris to grass clippings and leaves. Some are settleable, some floats and some are suspended; some are organic and some inert. As a result there is much variability in the way that stormwater solids are defined and characterized by different researchers, making it difficult to reconcile the findings of one study with another, and for regulatory agencies to develop meaningful regulations for solids management in urban runoff. Therefore, it is imperative to have consistent definitions and monitoring procedures to improve current stormwater management practices. The following is a proposed classification system that allows for consistent definitions of solids based on size and organic content. This classification scheme takes into account the environmental endpoints, practicality of sampling methods, and treatability.

To begin, a reasonable approach is to divide stormwater solids into four major size classifications. *Gross solids* include trash, litter, debris, and gravel sized sediment that travel as floating, suspended, or bedload in stormwater. These larger solids degrade aquatic habitat, smother productive sediments, leach harmful pollutants into the water column, and are an aesthetic problem (Rushton and England, 2006). Aquatic organisms can become entangled or ingest gross solids causing fatalities. *Coarse solids* are larger particles that travel in suspension or as bedload, depending on their specific gravity. Sediment deposition may alter spawning habitats, making them unsuitable for fish to lay eggs. Deposited particles may obscure sources of food, habitat, hiding places, and nesting sites (Wilber, 1983). However, the presence of coarse solids in urban stream may contribute positively to geomorphic stability, since they are less susceptible to scour than fine solids. *Fine solids* travel in suspension, but are typically settleable depending upon their density and size. Fine solids and coarse solids are often attributed to transporting harmful pollutants which can potentially bioaccumulate or cause chronic problems in organisms. In addition, fine solids degrade habitats by increasing turbidity which reduces light penetration. This negatively impacts photosynthetic organisms and can affect predator-prey relationships by decreasing visual abilities. Fine solids can also cause gill clogging, choke filter mechanisms on filter-feeding invertebrates, and clog feeding mechanisms of some zooplankton (McCabe and Sandretto, 1985). Finally, fine and coarse solids can infill spaces between larger solids in river beds needed for habitat (ASCE, 1992; Lenat et al., 1981; Lenat, 1984; Walters, 1995, Snodgrass et al., 1997, Simons and Senturk, 1991) resulting in less diverse aquatic populations (U.S. EPA, 2003). Finally, solids that are classified as *dissolved solids* comprise fine clays, colloidal materials, microorganisms, and bacteria in addition to dissolved chemicals. Thus, solids that are classified as dissolved are not truly dissolved in the water column. It is generally difficult to remove dissolved solids from stormwater runoff using BMPs relying on filtration or sedimentation.

Physical separation can be used to measure the size fractions of each of these four classes of solids in a stormwater sample. These solids can then be further classified as settleable and non-settleable, and volatile or non-volatile to further identify the impacts the solids have on receiving waters.

Gross solids are defined as the solid material that can be captured on a 5 mm screen. The No. 4 sieve in the US standard sieve size corresponds to 4.75 mm (close to 5 mm) and represents the separation between coarse sand and gravel (ASTM Standard D 2487-92). The size also is appropriate for sampling gross solids in the field using a net or screening device. Finally, visual separation can be performed on solids larger than 5 mm if the monitoring program requires it. A 5 mm size classification is consistent with studies in Australia (Allison et al., 1998) and California (Sullivan, 2005) on gross solids. Gross solids can further be divided into three classifications; litter, debris, and coarse sediment. Litter includes human derived trash, such as paper, plastic, Styrofoam, metal, and glass. Debris consists of organic material including leaves, branches, seeds, twigs, and grass clippings. Coarse sediments are inorganic breakdown of soils, pavement and building material (Rushton et al., 2006).

Coarse solids are defined as the solid material greater than 75 μm and less than 5 mm. These solids are associated with sedimentation destroying habitat, smothering benthic organisms, and transporting toxic elements into the ecosystem. Often, particles larger than 75 μm are not effectively collected using automatic water quality samplers therefore a combination of bedload samplers and autosamplers may be needed to sample this size range. The No. 200 mesh in the U.S. standard sieve size corresponds with 75 μm and is considered the separation between clay and silt and fine sand (ASTM Standard D 2487-92).

Fine solids are defined in this study as the material that passes through the No. 200 sieve (75 μm) but is retained by a 2 micron filter (Whatman grade 934AH or equivalent). Fine solids are commonly transported as suspended solids and attributed to increased turbidity, transporting harmful toxins into the ecosystem, and embeddedness characteristics.

Standard Methods for the Examination of Water and Wastewater allows several sizes with 2.0 μm as the maximum size. For standardization purposes, it is recommended that the 2 μm filter be selected because smaller size filters tend to clog and the residue itself affects the size of material that is retained above the filter. In addition, the 2 μm particle size represents the lower limit of particle size that will normally settle out in a typical stormwater detention pond. The distance solids will settle in 40 hours is shown in Figure 7-1 using Stokes Law with the following assumptions.

1) Water temperature is 8°C.
2) The absolute water viscosity is constant at a value of 0.014 g cm^{-1} s^{-1} (1.4 cP)
3) The density of water is constant at a value of 1.0 g cm^{-3}
4) The particles are spherical, therefore $\alpha = 1.0$
5) The predominant forces acting on the sphere is the buoyancy force and the force due to gravity. This does not take into account turbidity.
6) The graph ignores hindered settling, coagulation, and flocculation.

The particles greater than 2 μm include silt and clay-sized particles. Often the size of clay is designated as the percent finer than 0.002 mm, although Unified Soil Classification System ASTM D 2487 defines clay as the material passing a 0.075 mm screen that exhibits plasticity and

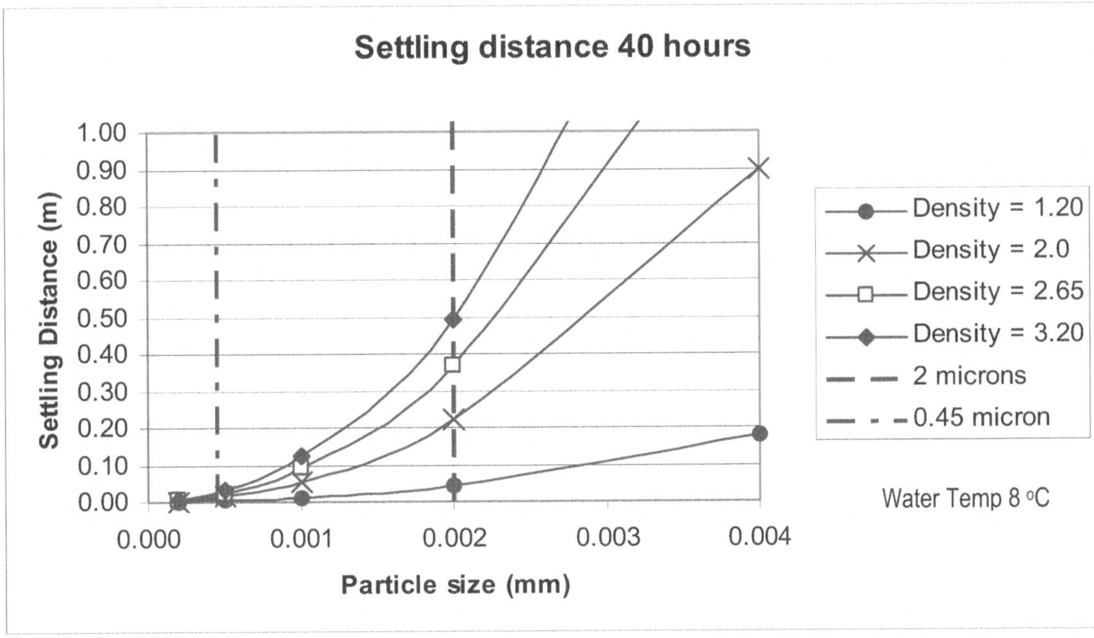

Figure 7-1. Settling Distance for Solids Removal in a Typical BMP Assuming 40-Hour Detention.

possesses strength when dry (Plasticity Limit greater than 4, PI vs. LL on or above "A" line). Silt is the material that passes a No. 200 sieve that is nonplastic or very slightly plastic and exhibits little or no strength when air dried (Plasticity Limit less than 4, plot of PI vs. LL is below "A" line) (ASTM D 2487).

Dissolved solids are defined as the particles that pass through a 2 micron filter (SM 2540 C). The proposed solids classification diagram and solids based on particle size is shown in the following Figures 7-2 and 7-3.

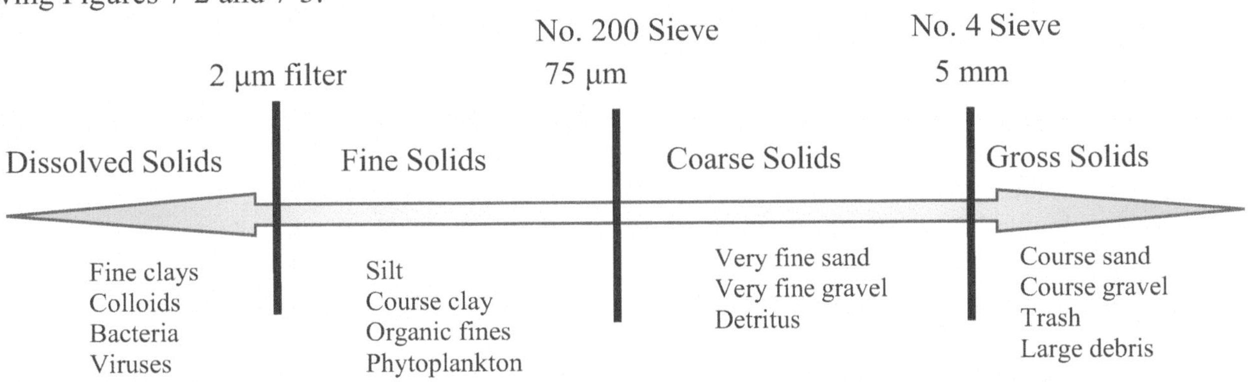

Figure 7-2. Solids Size Classification Diagram.

Figure 7-3 shows the types of solids that would be captured in the proposed solids classification. This figure includes United States Department of Agriculture (USDA) and International Society of Soil Sciences (ISSS) soil definitions. Some of the finest solids would be comprised of bacteria and its substrates.

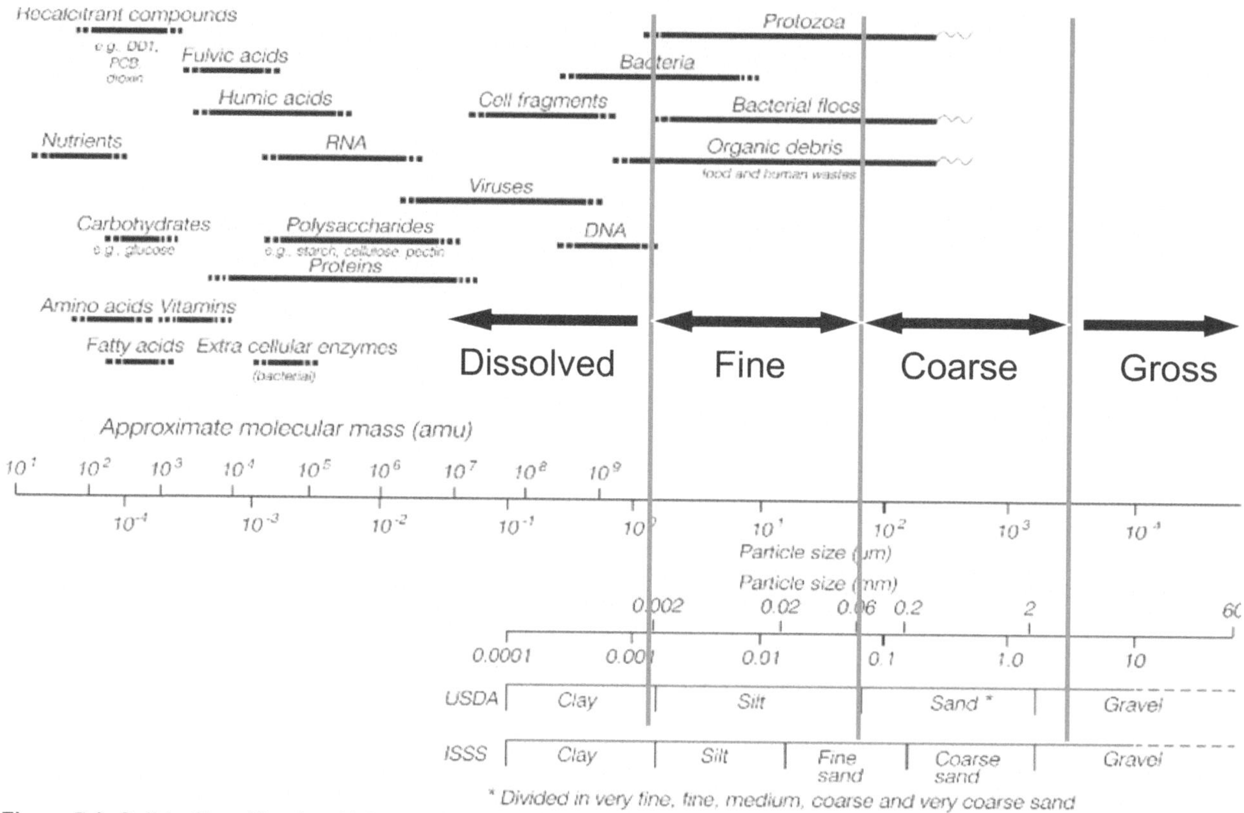

Figure 7-3. Solids Classification (Adapted from Ashley and others, 2004. Originally adapted from Levine A.D., Tchobanoglous, G. and Asano,T. 1985. Characterization of size distribution of contaminants in wastewater: treatment and reuse implications. J. WPCF, 57, 805-816)

The above solids classifications can further be divided into volatile (relating to organic) and non-volatile (relating to inorganic). Organic material comes from a wide range of sources including anthropogenic and natural sources. Examples of anthropogenic sources include paper, plastics, oil and grease, tire rubber, power plant, incinerators and pesticides and herbicides. Natural sources include organisms, leaves, grass, flower shatter, and pollen. Some examples of organics that are particularly harmful in the environment include oils, hydrocarbons, pesticides and herbicides. Additional negative impacts that increased organics may have are increased oxygen demand, phosphorus, ammonia, nitrite, and nitrate. Also, organics can serve as a substrate for bacterial growth. Increased bacteria and viruses in water can potentially be harmful to humans and aquatic life causing illness.

The relationships of these proposed definitions for Gross, Coarse, Fine, and Dissolved solids to receiving water impacts are illustrated in Figure 7-4.

Figure 7-4. Proposed Solids Classification and Associated Impacts.

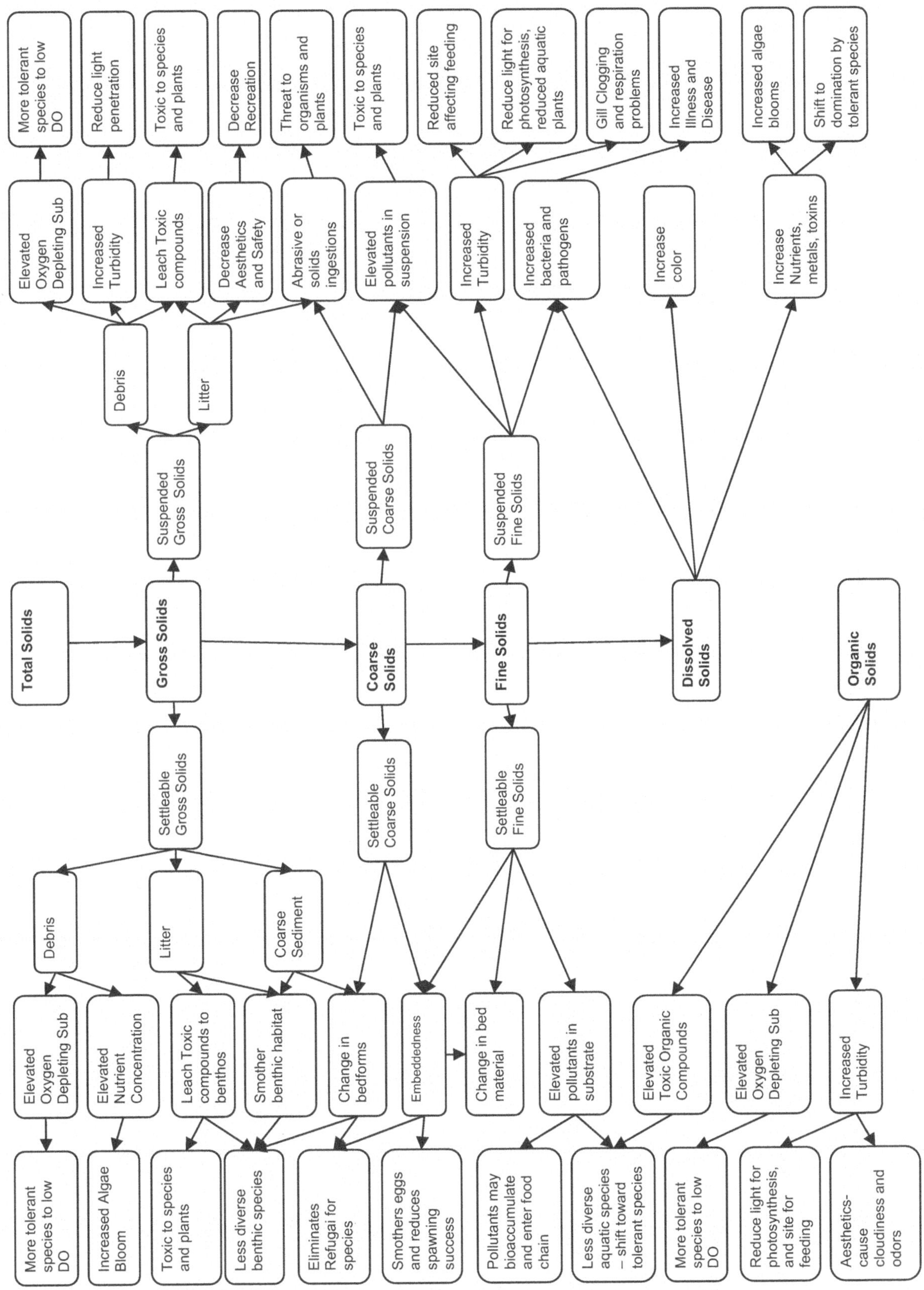

CHAPTER 8.0

PROPOSED STORMWATER SOLIDS ANALYSIS

8.1 Protocol for Analysis of Stormwater Solids

The solids types and concentrations in stormwater are highly variable, differing in size, density, and composition, both temporally and spatially, depending on many factors including rainfall intensity and duration, runoff peaks and volume, topography, particle characteristics, etc. Factors that affect the type and size of solids transported in urban runoff include intensity and duration of rainfall, duration of dry periods between storm events, and human activities within the catchment area (including litter, pet droppings, lawn care, traffic density, road maintenance, spills etc.). Other factors include catchment shape, land use, soil type, surface type and condition, etc. It is not surprising that there is not one commonly accepted size or density of particle that is problematic in stormwater. For example, Murray and others (1999) found the medium sand fraction to contribute the largest percentage to the total metal concentration in samples in the Detroit area (1999). In contrast, Sartor and Boyd (1972) found that less than 10% of particulates are in the silt and clay soil size but they contain over half the phosphorus and 25% of other pollutants. These results support the recommendation for performing more than one solids test on water samples.

There are many methods and analytical tests for classifying stormwater solids. Figure 8-1 shows the proposed analytical system for stormwater-borne solids based on this research. It is recommended to choose from these analytical tests the ones that will best suit the objectives of a given stormwater management program by targeting the solids of concern in the watershed and developing a detailed plan for sampling and analyzing the solids. The following sections provide guidance and recommendations for analytical determinations of the various types of solids identified in Figure 8-1

8.2 Gross Solids

Each site is unique and the decision about which form of gross solids to analyze should be based on the purpose and goals of the monitoring program. The choice can range from minimal laboratory analysis to highly time consuming characterizations (Rushton et al., 2006). Guidelines for developing a gross solids analytical plan are outlined in this document, although a well thought out program should be prepared that is consistent with the stormwater management program goals. As with any analytical plan, it is important to keep accurate records throughout the monitoring and analysis program.

The basic level of evaluation should include a volume measurement of gross solids collected and the time between cleanouts. Volume can be characterized in the gross solids trapping device by taking a depth measurement and multiplying it by the area. Alternative methods to calculate volume are to loosely place the gross solids in containers of known volume or record the volume occupied in the vacuum or haul truck needed at the time of cleanout.

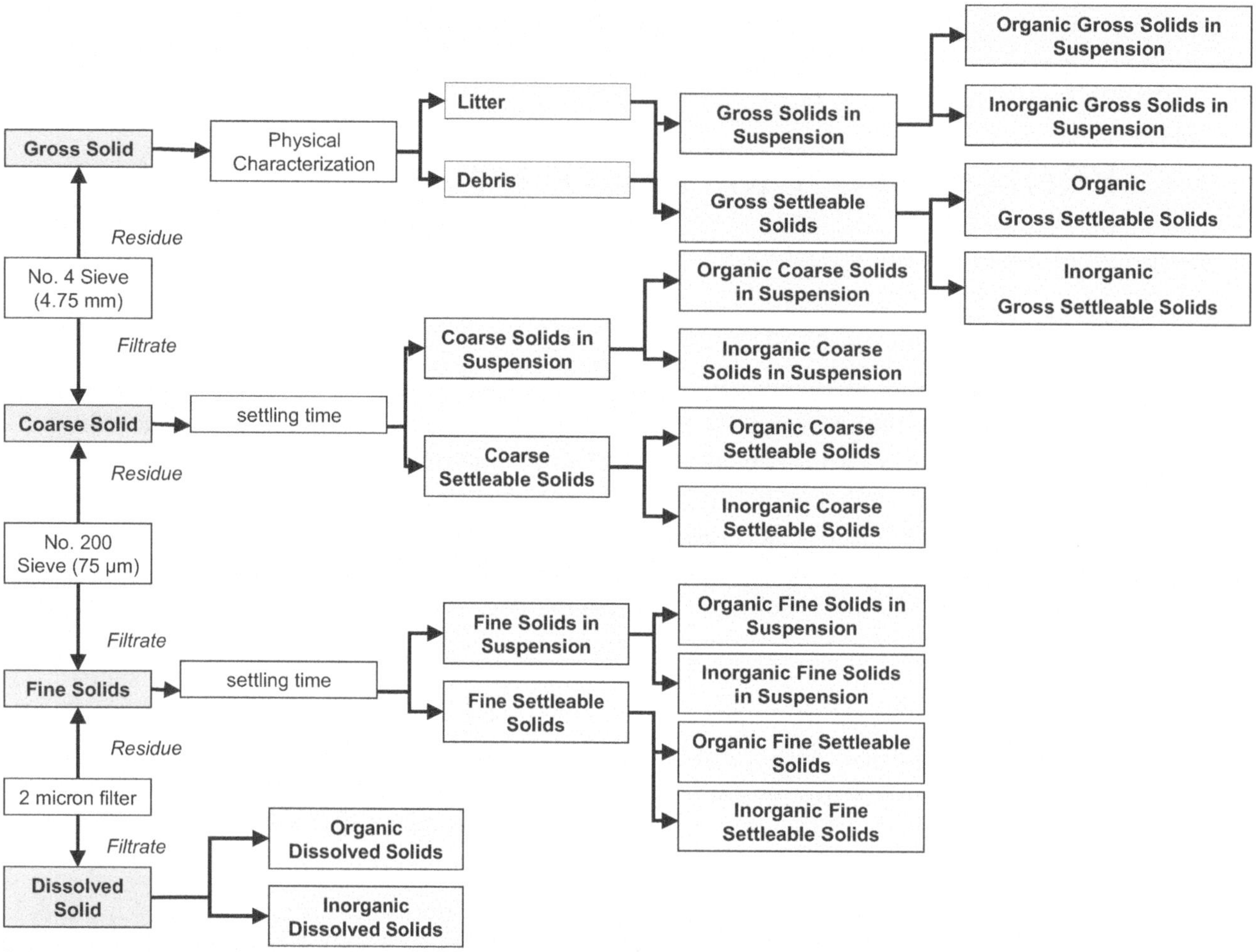

Figure 8-1. Solids Analytical Classification Diagram.

Depending on the goals of the monitoring program, more detailed gross solids characterization may be needed. Separation of litter from sediment and debris can be done by hand. Debris consists of natural organic material including leaves, branches, seeds, twigs and grass clippings (Rushton et al., 2006). Litter can be further categorized by visual separation into the groups indicated below. Characterizing litter may help in understanding the source of the litter and implementing pollution prevention plans.

- Cardboard/chipboard
- Paper
- Plastic
- Glass
- Styrofoam / Polystyrene
- Metal
- Wood - processed
- Cloth / fabric
- Cigarette butts
- "Other"

Mass and volume of the separated material can be determined. Litter can be characterized by weight, volume, and number of items; air-dried weight obtained using a digital scale; volume estimated by placing the litter samples into graduated containers; and the number of items determined by manual count. Air-dried weight can vary depending on humidity, temperature, and drying time. It is recommended that if weights are recorded to keep an accurate record of how the material was dried and report it with the data.

The percent volitile organic matter of the gross solids can be determined using standard methods by igniting the solids at a constant 550° C (SM 2540 E) where the solids lost upon ignition represent the volatile (organic) matter. This procedure does not distinguish precisely between inorganic and organic matter because the loss on ignition is not strictly confined to organic matter, it may include losses due to decomposition or volatilization of some mineral salts. If more precise characterization of organic matter is desired, tests such as total organic carbon (SM 5310), biochemical oxygen demand (SM 5210), and chemical oxygen demand (SM 5220) could be performed.

With an increase in program level and desired output detail, there will be a corresponding increase of effort and expense to more accurately separate the organic particles from the coarse sediment and determine the associated pollutants of each characterization. The chemical analysis and intensity of the study should be determined and a customized program designed to monitor targeted gross solids. Separation of sediment from debris and litter is time consuming and labor intensive. A guideline for possible gross solid analysis is described below.

1. Volume - Determine the appropriate volume and mass of the sample to ensure a representative sample recognizing that very large samples may be needed to represent the gross solids that persist in the catchment area. Volume estimations should be made at the time of collection and/or cleanout for the gross solids collected.
2. Characterization – Estimate the percent litter and debris. If desired, manually separate the litter (human derived trash) from organic debris and coarse sediment. More intense separation should be performed depending on monitoring goals.
3. Mass - Record the mass of the solids collected. Keep accurate records on how the solids were dried, the length of drying time, temperature, and humidity.
4. Volatile Solids - Ignite solids at 550°C for one hour in a muffle furnace using procedures described by standard methods. The difference in weight before and after combustion represents the fraction of solids that are volatile.

More intense studies may be needed to determine the percent of material that will float or the chemicals that will potentially leach into the water. If metals or other inorganic compounds are being tested, sampling and handling equipment should be suspended over a tub and rinsed from the top down with 10 percent nitric acid using a pump or squirt bottle (Rushton, et al., 2006; ASTM, 2000).

Gross Solids analyzed should be reported as a mass (kg) and/or volume (m^3) over time, while the pollutants associated with Gross Solids should be reported as mg/kg. The time between collections should also be noted when reporting the mass of solids accumulated between cleanout events (kg/year). If the study includes separation of litter and debris, mass and volume of each category should be recorded.

8.3 Suspended Solids

The traditional method for determining suspended solids in a water-solids mixture is the Total Suspended Solid (TSS) test (EPA Method #160.2 or Standard Methods 2540D). This method was originally developed to quantify solids with respect to wastewater treatment efficiency and thus can be expected to exhibit deficiencies when applied to stormwater, and it may not be suitable for natural waters (Gray et al., 2000). A modification of the test called Suspended Sediment Concentration (SSC), uses a similar laboratory method (ASTM Method D 3977-97), but the entire sample is filtered and analyzed. This procedure reduces the variability of the error associated with aliquot subsampling. *However, because of the legacy standing of TSS as a controlled pollutant and the limitations of the SSC test itself in measuring the class of solids of concern in stormwater quality management, it is not recommended that other tests be substituted for the parameters, but rather that the laboratory analysis of TSS be modified to more accurately reflect the solids fraction in stormwater that the test was meant to characterize in wastewater.* Refer to Chapter 10 for further discussion on the regulatory issues associated with TSS and SSC.

Neither the TSS nor SSC analytical procedures address the settleability potential of the solids, which is important in describing the transport potential and fate of pollutants. Understanding how the particles settle may provide better understanding how they will be transported, their ecologic impact on receiving waters, and their treatability. The ability for a particle to settle is a function of the size, shape, density, and surface charge of the particle. Therefore, an analytical procedure is proposed herein to address the settleability of the solids by allowing a settling period to separate the total solids in the sample into a "settleable" fraction and a "suspended" fraction (determined using a modified TSS analytical procedure).

Many particles greater than 75 μm have low densities, especially organic material. These particles will remain in suspension or float and add to the suspended solids. *It is therefore recommended to include these particles (greater than 75μm, but less than 4.75 mm) in suspended solids testing.* The separation of gross solids from coarse and fine solids should be done by sieving the entire sample through a No. 4 sieve (4.75mm). The solids that are retained on this sieve should not be included in the suspended solids analysis. If the separation between coarse and fine solids is desired in the monitoring goals, additional separations are required and described in further detail later.

8.3.1 Total Solids in Suspension (TSiS) with Mixing as an Alternate to SSC

Inconsistencies in the current methods to analyze TSS include mixing speed, pipette location, and pipette size. The TSS and settleable solids test were originally designed to analyze wastewater. Organic solids in wastewater typically have a specific gravity near 1.0, therefore requiring little agitation to become completely mixed. Particle density in urban runoff is much higher, ranging from 1.0-2.86 (Karamalegos et al., 2005), and in most cases, stormwater particulates have specific gravities in the range of 1.5 to 2.5 (Pitt, 1979). These particles with higher densities require more aggressive mixing to prevent solids settling. If the mixing speed is too slow the larger solids will readily settle out and be concentrated at the bottom of the beaker. The Standard Methods for Examining Water and Wastewater does not specify a mixing speed or pipette size before taking the aliquot.

The Standard Methods protocol for TSS has several deficiencies. There are variations in pipette sampling point within the sample, sample mixing speed is not sufficient to keep heavier solids suspended, and the pipette orifice size limits the size of solids that can be sampled. If the subsample is withdrawn using a pipette that is too small it may clog and prevent a representative amount of solids to be sample. In contrast, if the pipette is too large, solids may settle out of the pipette when the energy of mixing is lost. These issues can markedly alter the total solids count recovered in the subsample. The current TSS method needs to be revised at least to include an identified mixing speed and pipette for standardization purposes.

Based on findings of this document, it is recommended that the sample be mixed with a magnetic stirrer at a speed of 600 rpm in order prevent settling. Prolonged mixing may result in a change of particle size, therefore it is recommended to mix for no longer than one minute before taking the aliquot. Stenstrom (2006) conducted an experiment with a known concentration of solids with varying particle size distribution. A TSS analysis was performed at mixing speeds of 200-1100 rpm. A comparison was made (% recovery) between the known concentration and the concentration obtained from the TSS analysis. It should be noted that as particle size increases, it is increasingly difficult to keep the solids well mixed. For example, solids with a diameter less than 100 μm have a higher percent recovery compared to the true solids in the water/solids mixture than solids with a diameter greater than 250 μm with the same density and mixing speed. The aliquot sampling should be performed with a wide bore pipette diameter or sewage pipette from *mid-depth* in the sample. The results of solids recovery at variable mixing speeds and pipette openings can be seen in Figure 8-2.

TSS aliquots collected from the upper section tend to be biased low, while aliquots collected from the lower section tend to be biased high because of settling and the inability to maintain a well mixed sample (Kayhanian et al., 2006). In a study to improve the method of suspended solids measurements, Kayhanian and others (2006) observed a lateral concentration gradient with higher TSS concentrations near the wall of the sample container and lower TSS concentrations near the vortex. It was concluded that the sample should be taken midway between the vortex and the wall of the container. While not as accurate as SSC, better mixing and a large mouth pipette dramatically improve recovery (Stenstrom et al., 2006, personal communication). Based on findings of this research, it is recommended that the TSS analysis be modified as follows:

- Mix the sample using a magnetic stirrer for one minute at 600 rpm
- Use a large bore pipette. *The tip opening of pipettes recommended for use is about 3 mm in diameter (Kimble-Contes, Inc., 2000).*
- Withdraw the aliquot from mid-depth and midway between the vortex and the wall of the container.

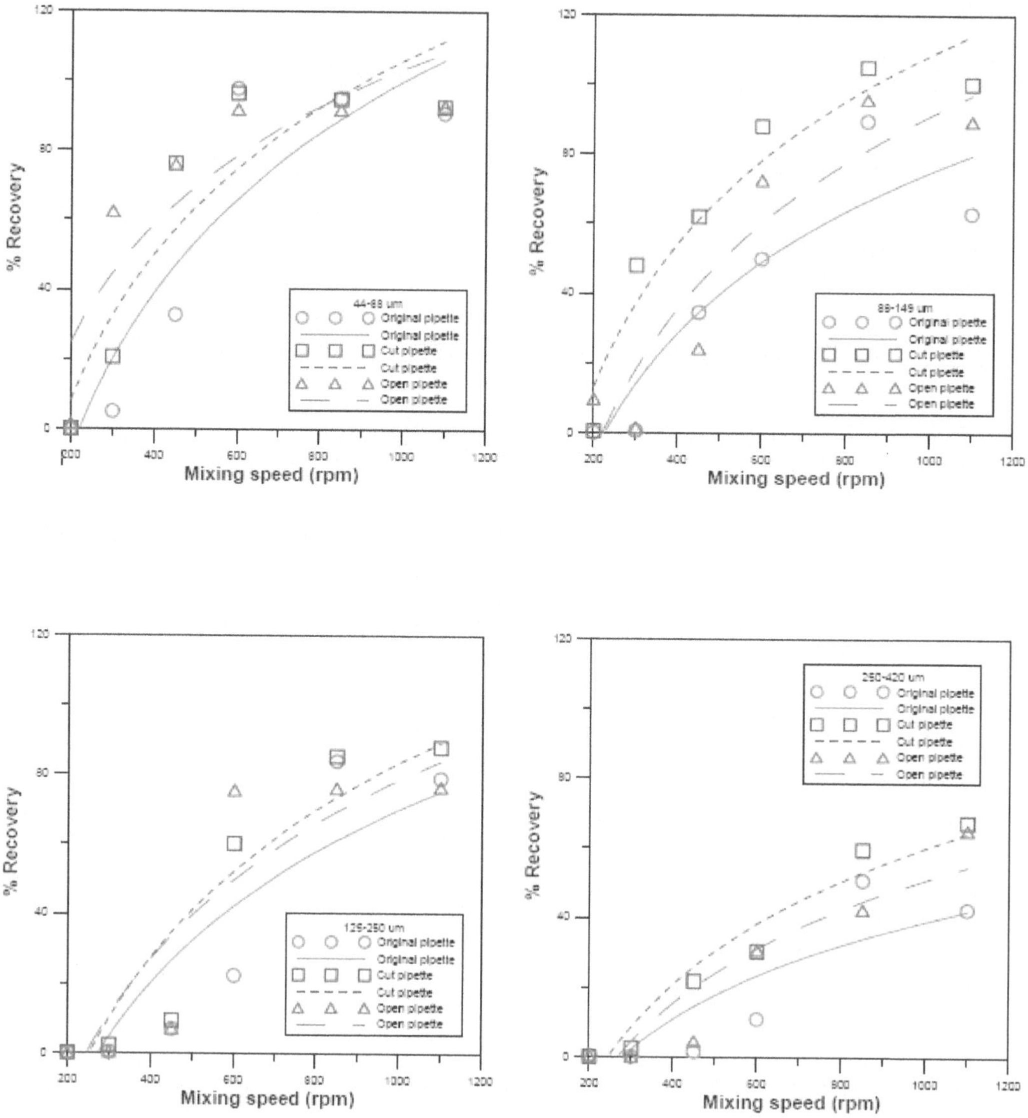

Figure 8-2. Solids Recovery with Variable Mixing Speed and Pipettes. (Courtesy of Michael Stenstrom, UCLA)

8.3.2 Total Suspended Solids with Settling as an Improved Estimate of the Traditional TSS Protocol

Settleable Solids (SM 2450 F) are currently measured by allowing not less than one liter of well-mixed sample to settle for one hour undisturbed. 250 ml is siphoned from the center of the vessel at a point halfway between the surface of the settled material and the liquid surface, and a TSS analysis is performed on this aliquot. The difference in concentration between the non-settled original aliquot and the siphoned supernatant aliquot yields the settleable solids concentration. The problem with this test is that the settling time (one hour) defines the particle size that will settle out and may not be appropriate for stormwater studies and removal practices. The settling velocity data for urban runoff indicate that a large fraction of TSS removed by sedimentation will settle out very rapidly (Driscoll, 1989).

Figure 8-3 shows a recommended method for solids testing that will result in the following parameters:

1. Suspended Sediment Concentration (SSC)
2. Total Suspended Solids (TSS)
3. Total Settleable Solids (TSTS)
4. Total Dissolved Solids (TDS)

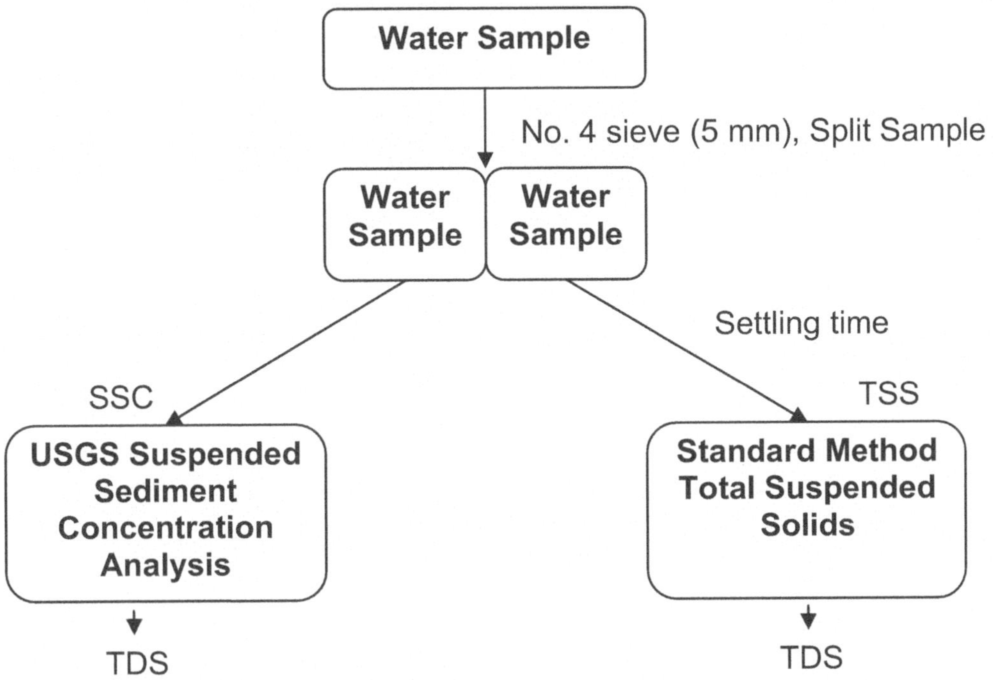

Figure 8-3 Recommended Suspended Solids Analytical Procedure.

The water sample that is obtained from the sampling device is sieved with a No. 4 sieve to remove gross solids, which could skew the analysis, and split into two equal samples using an appropriate churn or cone splitter. Care should be taken to obtain two equally representative samples. One water sample is then analyzed using ASTM Method D 3977-97 to determine the SSC, which represents all the solids greater than 2 μm for the entire sample (both fine solids and coarse solids). The procedure for determining SSC is outlined below (Method D 3977-97).

1. Place a 2 μm glass fiber filter on the membrane filter apparatus or insert into bottom of a suitable Gooch crucible with wrinkled surface up. While vacuum is applied, wash the disc with three successive 20mL volumes of distilled water. Remove all traces of water. Dry in an oven at 105 ± 1°C for one hour. Weigh the filter immediately before use. (For volatile solids, ignite for one hour in a 550°C oven)
2. Filter the entire sample of known volume through the 2 micron filter paper.
3. Rinse the sample container with DI water and pour the water onto the filter paper.
4. Dry the solids retained on the filter at 104 ± 1°C to constant weight.
5. Determine the mass of solids retained on the filter.

This concentration should be reported as the SSC in mg/l.

The second sample, for TSS determination, should be well mixed (following the procedure described in Section 8.3.1 above) and then allowed to sit undisturbed for five minutes to permit particles to settle or float out of suspension. A 5-minute settling time will allow for coarse separation and represents BMP processes relying on settling. Larger particles, with the same density, will fall at a faster rate than fine particles. As mentioned previously, fine particles are attributed with a greater negative impact on receiving waters due to increase specific surface area and the impacts fine solids have on ecology. A study by Contech Stormwater Solutions (2006) analyzed two water solids mixtures for TSS concentrations with settling times of 0-5 minutes. The results showed that there is a reduction of TSS with increasing time, but the rate of reduction begins to level out between 4-5 minutes. This indicates that the rapidly settleable solids have separated and that the solids that will remain in suspension are being sampled. Figure 8-4 shows a graphical representation of the experiment results.

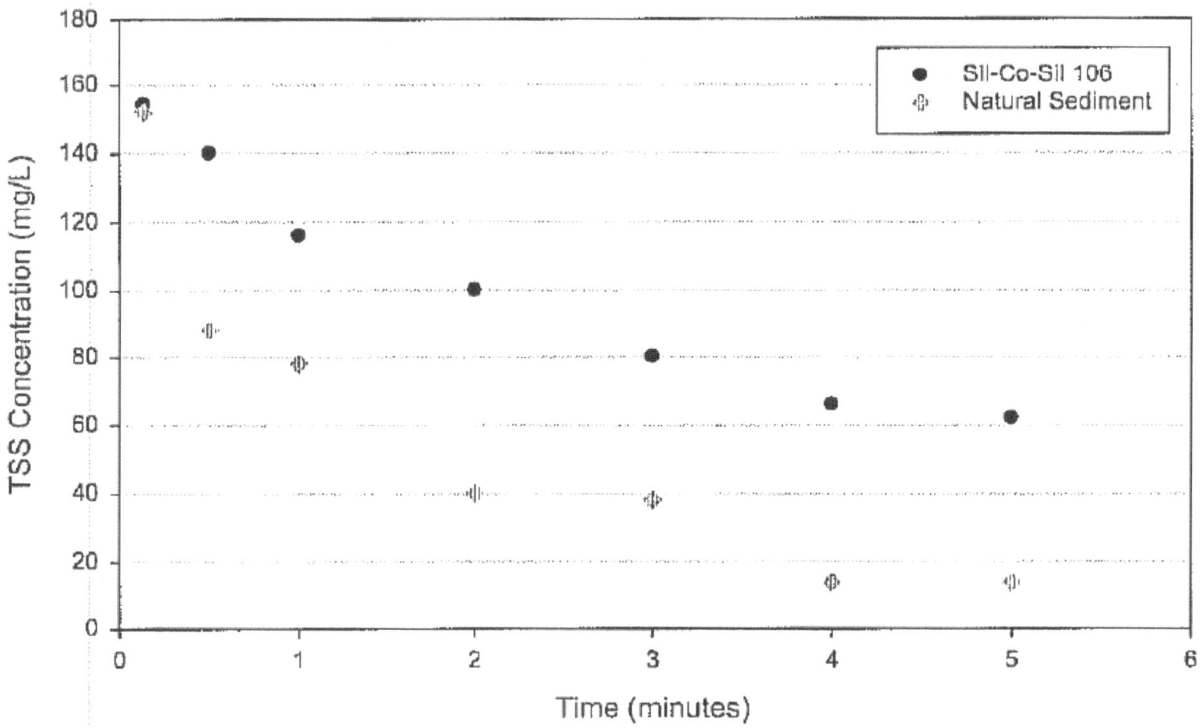

Figure 8-4. Graphical Representation of Allowing a Settling Time prior to the TSS Analysis. From Lenhart, James. Contech Stormwater Solutions. 2006

Many stormwater BMPs rely on settling for removal of solids in stormwater therefore the particle settling velocity should be investigated in the laboratory. To address the particles ability to settle out of suspension a short settling time is proposed to allow for separation. The specific gravity of discrete and agglomerate particles in stormwater is influenced by the presence of organic matter, which has specific gravities typically ranging from 1.1 to 1.5 (Minton, 2005). The solids in a study by Karamalegos and others (2005) showed that the densities of stormwater solids were highly variable, but it was concluded that the particle density in runoff ranged from 1.00-2.86 g/cm^3. These differences in densities associated with urban runoff affects the ability of the particle to settle out of suspension.

Figure 8-5 shows the settling distance of particles in 5 minutes. Heavier soil particles, such as sand and gravel, settle more quickly than finer silt and clay particles. The graph was created using Stokes law for several densities including 1.2, 2.0, 2.65, and 3.2 g/cm^3. There were several assumptions made which are listed below:

1) Water temperature is 8°C
2) Water viscosity is constant at a value of 0.014 g cm^{-1} s^{-1}
3) The density of water is constant at a value of 1.0 g cm^{-3}
4) The particles are spherical, therefore $\alpha = 1.0$
5) The predominant forces acting on the sphere is the buoyancy force and the force due to gravity. This does not take into account turbidity
6) The analysis ignores hindered settling, coagulation and flocculation

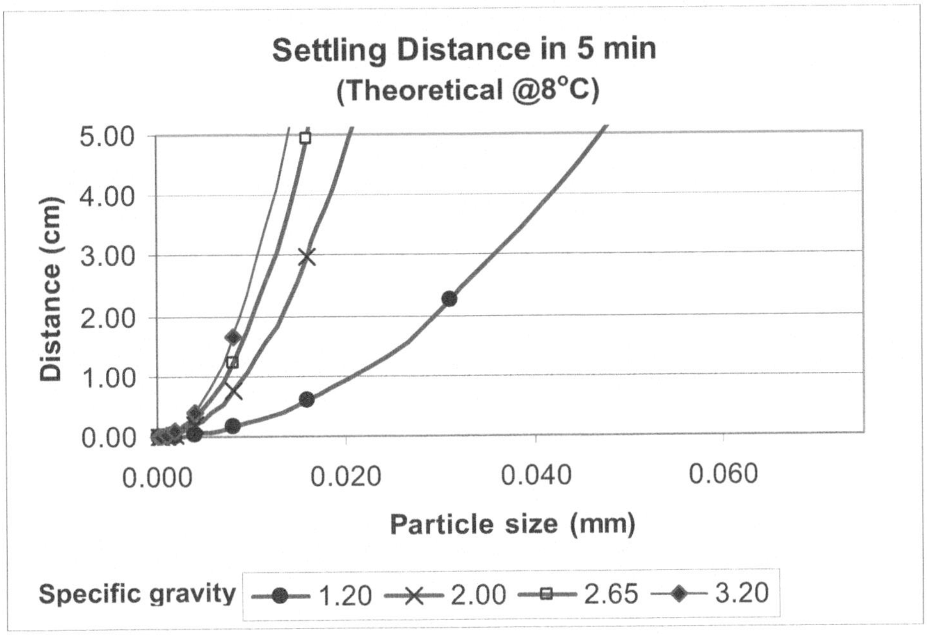

Figure 8-5 Settleable Solids Analysis According to Stoke's Law.

As temperature of the water increases the viscosity of the water will decrease which will make a difference in settling velocities. In addition, the density of water will increase with increasing dissolved solids which profoundly affects settling rates in regions where road salting is a common de-icing practice.

The proposed procedure to determine the TSS and the Total Settleable Solids is described below. Essentially, it is the same procedure as the current TSS method with a settling period before the withdrawal of the subsample. Initial preparation to the filter, evaporation dish, and sample should be the same as identified in current Standard Methods.

1. Prepare a 2 μm glass fiber filter disk according to standard methods (Standard Methods for the Examination of Water and Wastewater 2540)
2. Place the glass fiber filter on the membrane filter apparatus or insert into bottom of a suitable Gooch crucible with wrinkled surface up. While vacuum is applied, wash the disc with three successive 20mL volumes of distilled water. Remove all traces of water. Dry in an oven at 103-105°C for one hour. Weigh the filter immediately before use.

3. Heat and clean an evaporation dish according to Standard Methods: Heat the clean evaporation dish to 103-105°C for one hour, if volatile residue is to be measured, heat at 550 ± 50°C for one hour in a muffle furnace. Cool, desiccate, weigh and store in desiccator until ready for use. Weigh immediately before use.
4. Filter the entire sample over a No. 4 sieve to separate the Gross solids from the coarse and fine solids.
5. Stir the sample using a magnetic stirrer for one minute at 600 rpm.
6. Allow the sample to sit without disturbance for 5 minutes.
7. Take a subsample of 250 ml from the middle of the sample at mid depth.
8. Filter the sample through the pre-prepared 2 μm glass fiber filter as described by standard methods.
9. Dry the residue and filter at 104°C ± 1°C to constant weight.
10. Cool in a desiccator and weigh.
11. Determine the mass of solids retained on the filter.
12. Report the concentrations of residue in units of mg/l as TSS. The difference between the SSC and the TSS is the Total Settleable Solids (mg/l).

If the monitoring goal is to identify the separation between Coarse and Fine solids, the proposed procedure is illustrated in Figure 8-5. In this case, the separation of Coarse solids from Fine solids requires an additional sieving step to separate the solids larger than a No. 200 sieve.

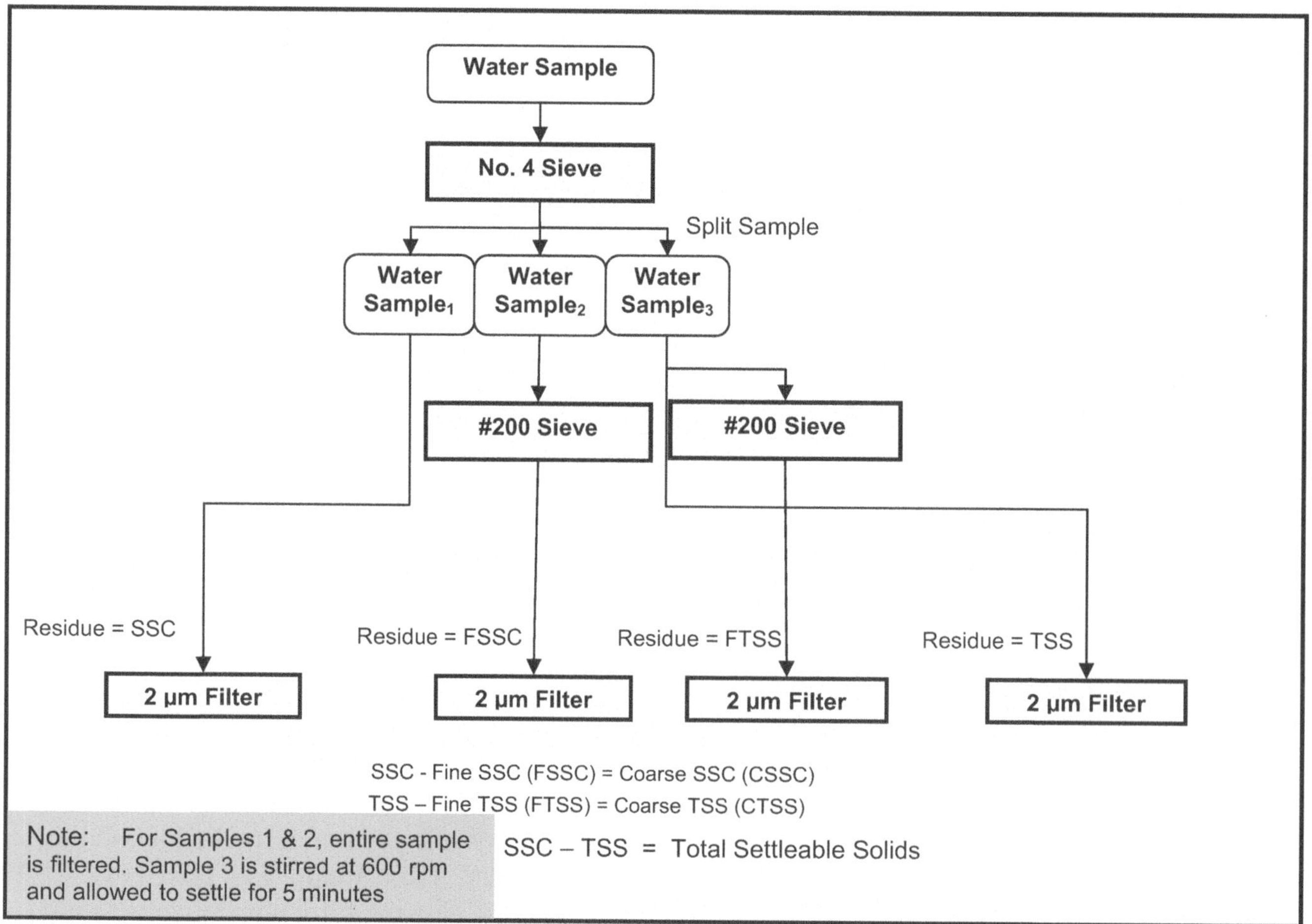

Figure 8-6. Fine and Gross Solids Analytical Procedure.

This separation includes an initial coarse separation in the USGS SSC method (ASTM Method D 3977-97) over a No. 200 sieve (75 μm) in order to determine the Coarse SSC and the fine SSC. In addition, a third measurement is done to determine the settleability of Fine solids. This measurement is done according to the modified TSS analysis protocol described above, except that after mixing and settling, coarse filtration of water sample 2 is performed over a No. 200 sieve. In this method, the types of solids measured include Coarse SSC, Fine SSC, TSS, and Fine TSS. From these values mass balance equations can be used to determine additional parameters including SSC, Coarse TSS, Coarse Settleable Solids, and Fine Settleable Solids using the difference between SSC and TSS values.

8.3.3 Suspended Solids using a Separatory Funnel

The SSC analytical procedure has been shown to be more robust than the TSS procedure (Gray et al., 2000) for analyzing total solids in suspension because the entire mass of solids is analyzed as opposed to a subsample. This test could be further developed to describe the particles ability to settle out of suspension. The following separatory funnel technique would permit the separation of solids that would settle out of suspension in a predetermined amount of time and the solids that remain in suspension. A diagram of a separatory funnel is shown in Figure 8-6. The outline of the proposed procedure follows:

1. Add entire sample to the separatory funnel
2. Gently invert the sample with stopcock and flask closed 5-6 times in 30 seconds
3. Allow the sample to sit undisturbed for 5 minutes
4. Remove the stopper and allow the lower portion of the sample to drain
5. Filter the top portion and the bottom portion separately through a 2 micron glass fiber filter and analyze following ASTM Method D 3977-97 for SSC.

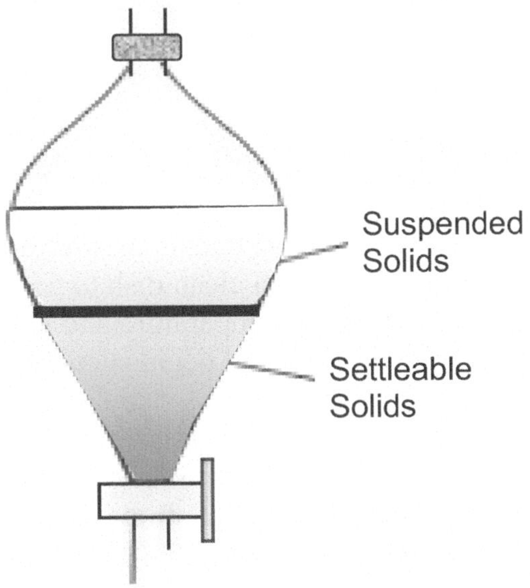

Figure 8-7. Separatory Funnel Diagram.

In this procedure the entire sample is analyzed (similar to the SSC analysis procedure), but the sample is allowed a settling period to separate the suspended solids from the settleable solids. The top layer and bottom layer together are equivalent to the SSC. If the analysis objective is to identify Course solids and Fine solids, an initial filtration over a No. 200 sieve is recommended to separate the two solids sizes.

Advantages

- The entire mass of solids is analyzed on a known volume of sample (similar to SSC). This analysis will reduce the error associated with subsampling using a pipette or pouring from an open container.
- TSS and Total Settleable Solids are addressed by allowing a sedimentation period. This distinction is useful in describing the transport and fate of the solids as well as describing the ecological impact of these solids of different sizes and densities.

Disadvantages

- The separatory funnel must have sufficient volume to hold the entire sample.
- Rate of removal of settled material from the outlet would have to be controlled, in order to avoid mixing effects.
- The outlet at the bottom of the funnel must be large enough to allow all particles that would be trapped using an automatic sampler to pass through. This limitation is not expected to be a problem.

8.4 Dissolved Solids

Dissolved solids should be tested following standard methods described by EPA Method 160.1 or Standard Methods for the Examination of Water and Wastewater 2540 C. The basic procedure is, after filtering the sample through a 2 μm filter, the filtrate is evaporated to dryness in a weighed dish and dried to constant weight at 180°C. The increase in dish weight represents the total dissolved solids. The procedure SM 2540 C is listed below:

1. Prepare a 2 micron glass fiber filter disk. Insert disk with wrinkled side up in filtration apparatus. Apply vacuum and wash disk with three successive 20-mL portions of reagent-grade water. Continue suction to remove all traces of water, turn vacuum off, and discard washings.
2. Prepare evaporating dishes. Heat a clean dish in an oven to 178-182°C for one hour. If volatile solids are to be measured, heat a clean dish to 550°C for one hour in a muffle furnace. Cool in desiccator to balance temperature and weigh immediately before use.
3. Assemble filtering apparatus and filter and begin suction. Wet filter with a small volume of reagent-grade water to seat it.
4. Pipette a measured volume onto the seated glass-fiber filter.
5. Wash filter with three successive 10-mL volumes of reagent-grade water, allowing complete drainage between washings, and continue suction for about 3 min after filtration is complete. (Samples with high dissolved solids may require additional washings.)
6. Transfer the filtrate to a weighed evaporating dish and evaporate to dryness
7. Dry for at least 1 hr at 180°C ± 1°C in an oven, cool in a desiccator to balance temperature, and weigh. Repeat the cycle of drying, cooling, desiccating, and weighing

until a constant weight is obtained or until the weight change is less than 4% of the previous weight or 0.5 mg, whichever is less.

Report the concentration as mg/L of Total Dissolved Solids. Most stormwater BMPs that rely on settling do not remove dissolved solids, although this test may provide some information on the very fine clays and colloids in the water which can transport harmful pollutants. Solids larger than about 10^{-4} μm require further disintegration before microbial uptake and utilization is possible (Ashley et al., 2004).

8.5 Organic Solids

The volatile solids test has good correlation to organic solids, is relatively simple and offers significant information about the impact of the solids on the receiving water. The standard method (U.S. EPA Method 160.4) involves drying to establish the total mass of the solid followed by ashing the sample in a muffle furnace at 550°C to burn off the organic matter. The total mass lost in this process represents the volatile organic matter. It is recommended to perform an organic solids test in addition to classifying the solids as gross, suspended, or dissolved. This test can be performed by adding the following two steps to the modified solids tests described above.

1. Dry the residue and filter at 550 ± 50°C in a muffle furnace for at least one hour to constant weight
2. Cool in a desiccator and weigh

The solids lost through ignition represent the volatile organic solids and can be reported as mg/l volatile solids. When this analysis is performed on a solids class size it should be reported as mg/l of volatile in the solids classification, for example volatile solids in suspension or volatile fine solids in suspension. As previously stated, the weight loss on ignition is not necessarily confined to organic matter because it may include losses due to decomposition or volatilization of some mineral salts. More precise characterization of organic matter may be made by such tests as total organic carbon, biochemical oxygen demand, and chemical oxygen demand if desired.

8.6 Particle Size Distribution

For more involved stormwater studies, particle size distribution (PSD) information is valuable for describing the transport and fate of the solids. It is strongly recommended that PSD analysis be performed on samples. There are several methods for determining PSD, including:

- Dry sieve
- Wet sieve
- Hydrometer
- Visual accumulation tube (VA)
- Pipette
- Light Scattering Counters
- Light Obscuration Counters
- Microscopy
- Coulter counter
- Sedigraph (x-ray sedimentation)

These methods were discussed in Section 5.3. An appropriate method should be chosen depending on the site specific parameters. It is cautioned that some of the PSD tests, such as microscopy, use small subsamples that may not be representative of the sample as a whole.

8.7 Chemical Analysis

When the goal of the monitoring includes site specific parameters, such as heavy metals or nutrient concentrations, these tests should be performed on the solids according to standard methods. Horner et al., (1994) describe in the Terrene Institute's *Fundamentals of Urban Runoff Management* several pollutants associated with urban runoff including solids, oxygen-demanding substances, nitrogen and phosphorus, pathogens, petroleum hydrocarbons, metals, and synthetic organics. The evaluation of pollutant specific tests was not included in this research but is encouraged if the goal of the monitoring is to assess the environmental impact of particular constituents.

When performing chemical analyses on metals in a water sample, a distinction between dissolved and particulate material is recommended. It is recommended that chemical analysis be performed on total recoverable metals to better understand the ecological impact. Dissolved species are generally more available to biota and to reaction with other components in the system than are particulates (Bricker et al., 2003). The standard practice is to define the dissolved fraction as the solids that pass through a 2 μm or smaller filter, although this is only an operational definition. Fine clay and colloidal solids will pass through the filter and carry with them metals and other toxins sorbed on their surfaces. Colloidal and truly dissolved species have different characteristics relative to bioavailability and toxicity (Bricker et al., 2003). A major finding of the Nationwide Urban Runoff Program was that metals, especially Cu, Pb, and Zn, are the most prevalent constituents found in urban runoff land uses such as transportation (U.S. EPA, 1983). It may be useful to monitor prevalent metals and toxins for these land uses.

CHAPTER 9.0

GUIDANCE FOR SAMPLING

9.1 Introduction

Stormwater sampling proves to be a difficult task; yet it is crucial for obtaining results that are truly representative of runoff quality. Stormwater quality tends to be extremely unpredictable (U.S. EPA, 1983; Driscoll et al., 1989). Therefore, obtaining a representative sample is a particularly difficult task because solids vary in size, concentration, time and location. The distribution of solids varies vertically as well as horizontally. It is also difficult to measure stormwater flow because of the uncertainty in precipitation and weather patterns. The irregular intensities of rainfall make it difficult to predict runoff rate, pollutant transport, sediment deposition and re-suspension, channel scour, etc. Pollutant sources such as landscape practices, spills, construction activities, traffic density, and vehicle washing drastically alter the pollutant load. It is not surprising that solids and pollutant characteristics can be different from location to location as well as from event to event.

There are several existing documents that outline stormwater sampling and monitoring methods that are listed in Table 9-1. These methods cover nearly every situation that might be

Table 9-1. Guidance Documents for Sampling Stormwater.

Organization	Title of Document	Website
United States Environmental Protection Agency and American Society of Civil Engineers	Urban Stormwater BMP Performance Monitoring (U.S. EPA-821-B-02-001) 2002	http://www.bmpdatabase.org
United States Geologic Survey	National Field Manual for the Collection of Water-Quality Data	http://water.usgs.gov/owq/FieldManual
United States Environmental Protection Agency	Stormwater Effects Handbook: A toolbox for Watershed Managers, Scientists, and Engineers	http://www.epa.gov/ednnrmrl/publications/books/handbook/index.htm
American Society of Civil Engineers, Urban Water Resources Research Council	ASCE Monitoring Guideline for Monitoring Stormwater Gross Solids	*Working document 2006*

encountered in the field, therefore no additional guidance is offered here. It is recommended that these resources be used as guidelines to develop a sampling plan to meet the specific goals of the water quality study that the sampling program is intended to support. A few of these sampling and handling guidelines are summarized below. It is recommended that the BMP Performance Monitoring document (U.S. EPA, 2002b) be used as a basic guideline, but the specific sampling methods used should be based on the study objectives and data quality objectives. In this regard, the information on Sample Collection and Handling in Chapter 4.0 may be useful.

9.2 Developing a Plan

Before sampling stormwater, it is necessary to develop a detailed plan. The development of a plan requires a high level of professional judgment and thoughtful consideration. The solids and pollutants of concern in stormwater are site specific. Likewise, the receiving waters and natural habitats that are affected by stormwater runoff are diverse. Therefore it is necessary to develop an individualized plan for the locality that addresses the specific problems in the area. For example, citizens of California were outraged over the amount of visible litter present in the waterways. Therefore the stormwater sampling plan targeted gross solids and items larger than 5 mm that could be seen and recognized. Their sampling data supported the development of a control plan to reduce Gross solids.

An outline by the U.S. EPA on Data Quality Objectives (1994) is listed below

- Determine project goals and objectives
- Identify resources and constraints
- Identify data characteristics and tools
- Determine key study parameters
- Specify methods for obtaining data
- Develop performance and acceptance criteria
- Optimize the design for obtaining the data

A few of the main considerations of the monitoring plan are site selection, number of monitored storm events and their temporal distribution, characteristics of target storm events, types of samples, and analytical constituents (BMP database EPA-821-B-02-001, 2002).

Accurate record keeping is important when monitoring solids. Some parameters to record include rainfall amount, intensity and duration, period since last rainfall, season, unusual weather events, and flow rate. Information about the land area is useful to record including watershed size, land use, geographical location, percent imperviousness, vegetative canopy, soil type, and watershed slopes. BMPs in the basin, a description of their design, and recorded maintenance should also be noted.

9.3 Special Considerations for Sampling Solids in Urban Runoff

This section outlines recommendations for sampling solids found in urban runoff (Wilde, 2005; U.S. EPA, 2002b). The sampling method used depends on the solids types of concern, which at dependent upon the receiving water impacts. Figure 9-1 should be useful for linking solids types of concern to impacts and sampling needs.

The particle size that will dominate the stormwater runoff depends on the surface soils in the area that the runoff is generated from, land use, and intensity and duration of the storm. Different regions have varying geologic and urban area conditions that tend to produce differing particle sizes in runoff (Landers, 2006). Some areas show sand size particles contributing the most to water quality degradation, while other studies show fine particles contributing the most to water

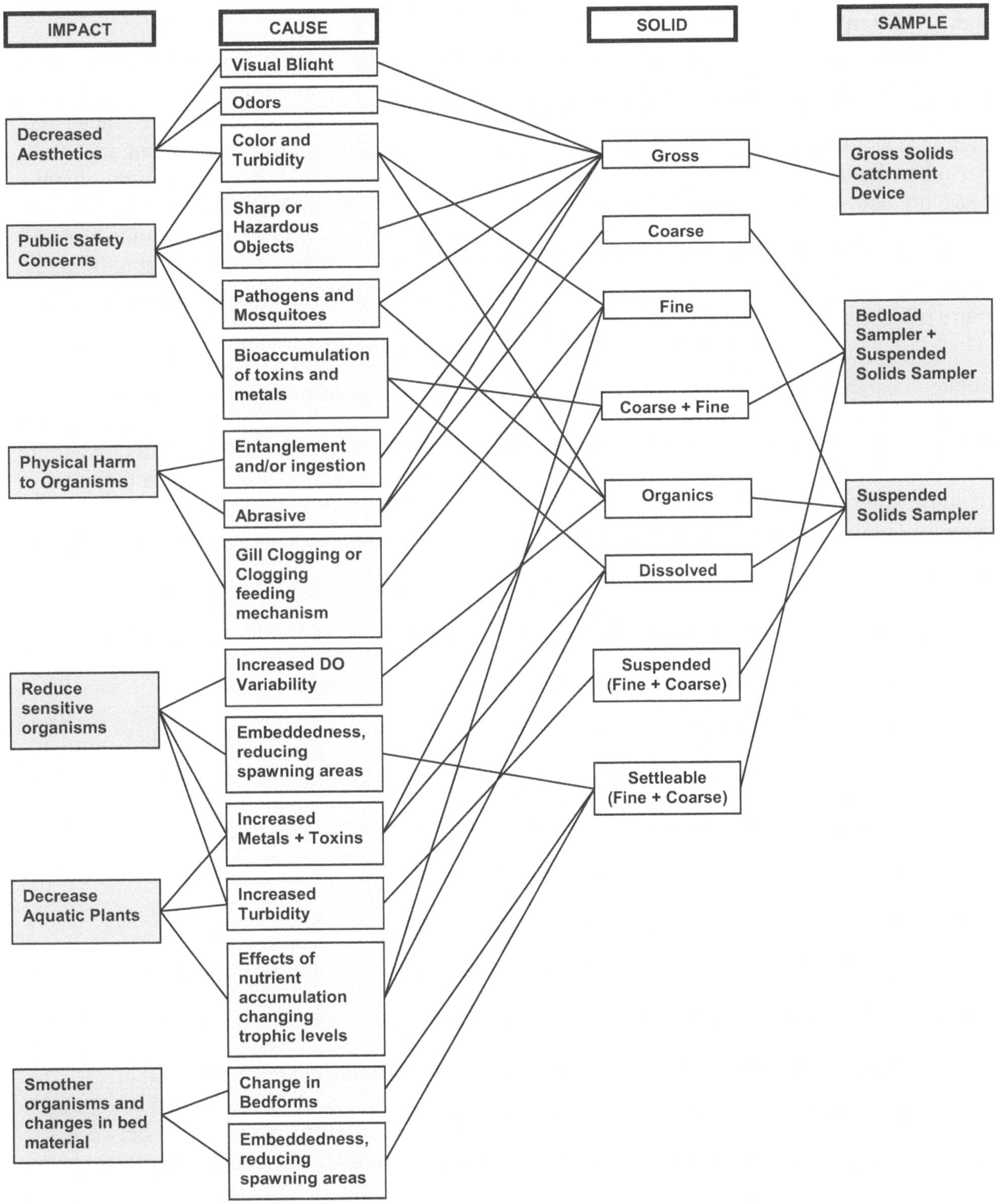

Figure 9-1. Sampling Solids Based on Impacts.

quality degradation. This means that no single test will sufficiently describe the water quality for all monitoring locations.

9.3.1 When to Sample

Site specific storm criteria should be developed to achieve monitoring goals. It is common for monitoring to begin by collecting samples under certain rainfall and/or flow conditions, although it is cautioned that many storms may fall outside of the specified range and therefore would not be sampled. If a "worst case" scenario is desired, long antecedent dry periods will most likely result in higher pollutant washoff because it allows for more pollutants to accumulate on the ground surface before being washed off. If total load is of interest samples should be taken throughout the duration of the storm. In order to obtain a representation of the solids transported throughout the storm it is necessary to take samples during the rising limb, peak, and falling limb of the runoff hydrograph. In most cases, the more samples obtained, the more representative the sample is of the system behavior.

It is important to know how many samples must be taken to be confident in the data obtained is representative and not "noisy" data. This section outlines methods to determine the number of samples needed to show statistically significant data (information is summarized from the Urban Stormwater BMP Performance Manual). This procedure assumes normality of sample data. Monitoring a large number of samples is required if the objective is to estimate the quality of stormwater in a given catchment. There are four factors that will impact the number of samples required to provide useful estimates of the mean and event mean concentration. These factors are:

1. Allowable level of error, or variance
2. Level of statistical confidence in estimates of the mean
3. Number of samples collected
4. Variability in population trends

The number of samples required is a function of the allowable error in the data means (E) and the standard deviation (σ). The equation below can be used to calculate the number of samples required when sampling at a single location.

$$n = \frac{4(\sigma)^2}{(E)^2}$$

Where: n = number of samples
σ = sample standard deviation
E = allowable error in the data set

Typical standard deviations can be estimated from existing data or from local or nationally published data on expected quality of stormwater runoff. A different method is used to calculate the number of samples required to show statistically significant difference between mean concentrations from two sample sets (such as significant difference between influent and effluent concentrations). The hypothesis test, or student t-test, can be used to evaluate whether the means of two data sets are statistically different by testing whether there is little overlap between the confidence intervals. Refer to the Urban Stormwater BMP Performance Monitoring Manual, Appendix C, for guidance on estimating the number of samples required to determine, with some degree of confidence, that observed means are statistically significant.

The t-statistic should be used in small sample sets (comprising less than 30 samples). The t-statistic depends on the number of measurements; therefore an iterative process is required. If the sampled data is composed of more than 30 samples the Z-statistic can be used in place of the t-statistic. The number of samples needed to achieve a desired confidence interval to show statistical difference in an influent and effluent sample is:

$$n = \left[\frac{(Z_{\alpha/2} + Z_{\beta/2}) \times COV \times (2 - \%removal)}{\%removal} \right]^2$$

$$n = 2\left[(Z_{1-\alpha} + Z_{1-\beta})/(\mu_1 - \mu_2)\right]^2 \sigma^2$$

Where:
- n = number of samples
- σ = sample standard deviation
- COV = coefficient of variation
- Z = statistic obtained from a standard normal distribution table
- μ = sample mean

The above equations assumes that the sample sets have the same number of samples (n), coefficient of variation (COV) and function of desired level of certainty ($Z_{\alpha/2}$ and $Z_{\beta/2}$). If the influent and effluent COVs are not the same, the following equation can be used to solve for the number of samples required:

$$n = Z_{\alpha/2} \left[\frac{COV_{in} + COV_{out}(1 - \%removal)}{\%removal} \right]^2$$

9.3.2 What to Sample

The parameters to monitor should be site specific and cater to the monitoring plan developed for the area. Stormwater can have many problem constituents that are responsible for degrading receiving water quality, causing immediate and long term stress to aquatic habitat, and potentially causing a direct threat to public safety. Pollutants in runoff can include herbicides and pesticides, toxic heavy metals, organic pollutants, bacteria and pathogens, sediment, excess nutrients, and litter. Sediment and sediment-related pollutants are the concentration of this research, although other parameters may need to be addressed depending on the monitoring goals. The solids that may be of concern sampled include:

- Gross solids – requiring a trapping device to retain solids greater than 5 mm.
- Coarse solids – often requiring bedload samplers to supplement autosamplers to sample solids that travel as bedload.
- Fine solids – sampled by autosamplers.

9.4 Fine Solids and Coarse Solids Sampling (Solids less than 5 mm)

In most urban settings, automatic samplers are the most suitable method for collecting samples of runoff (Bent et al., 2001). The automatic sampler intake should be placed at a point at which the concentration approximates the mean sediment concentration for the cross section. There are several resources that give guidelines on placing automated sampler intakes. (FHWA Publication No. FHWA-EP-03-054 2003, USGS National Field Manual for the Collection of Water-Quality Data) (Also see Chapter 4.0). Field and others (1997) advise that in order to determine suspended solids removal a sample must be taken across the entire cross section and vertically through the water column of the flow. The sampling system must have intake velocities greater than the mainstream velocity to draw heavier particles and multileveled ports to capture stratified particles. It is recommended to use automatic flow-weighted samples for monitoring efforts targeting fine and dissolved solids (See Section 4.2 on autosamplers).

It is known that autosamplers do not sample the largest particles that are found in urban runoff or solids that travel as bedload because of the size and location of the intake. Solids can travel in suspension, near bedload, or as bedload. In the case where coarse solids are of concern and more accurate particle size distribution is desired, it is recommended that bedload samplers be used to supplement automatic water samplers (Burton and Pitt, 2002). Bedload comprising coarse solids and can account for about 5 - 10% of the sediment load under some conditions and require special sampling (Burton and Pitt, 2002). Appropriate equipment should be used when sampling bedload and caution should be taken to minimize disrupting or resuspending solids when sampling. Most fluvial bedload samplers are appropriate to sample stormwater (Bent et al., 2001). Floatable solids larger than 5 mm may be captured using a net or screen.

9.4.1 Where to Sample

A few general sampling rules are listed below; although there are several resources with guidelines on stormwater sampling that should be referenced prior to adopting a sampling plan. The EPA and ASCE's Urban Water Resources Research Council have developed a guidance manual, *Urban Stormwater BMP Performance Monitoring* (U.S. EPA, 2002b) that outlines sampling guidelines including location of samples, equipment required, and techniques of sampling. The *USGS National Field Manual for the Collection of Water-Quality Data* (Book 9 chapter A1-A9) and the *NPDES Stormwater Sampling Guidance Document* (EPA 833-B-92-001) provides additional guidance for sampling and analyzing stormwater.

The number of samples and sampling location should be based on the goals of the monitoring program, the size of the drainage basin(s), and the resources available. When the objective of the monitoring is to determine spatial variations of urban runoff more locations with less frequent monitoring may be better. If the goal is to investigate the temporal variations, then fewer locations with more frequent monitoring may be more beneficial. Obviously, the more samples collected, the more information about the solids will be available. Urban Stormwater BMP Performance Monitoring Section 3 provides guidance to calculate how many samples are necessary to determine with some confidence that the observed means are statistically significant.

The most important objective when choosing a sampling location is to obtain a representative sample. The USGS defines a representative water sample as one that typifies ("represents") in time and space that part of the aqueous system to be studied and is delineated by the objectives and scope of the study. It is important to get a well mixed sample that is representative of the solids both horizontally and vertically. Some guidelines for choosing an appropriate site are listed below.

Monitoring in Storm Drain Systems

- The catchment area for the storm drain system should be well understood. Investigation is needed to confirm if the contributing catchment is completely served by a separate storm drain or by a combined sewer system. This will impact the contaminants in the effluent.

- Take samples that are well mixed and approach uniform flow conditions. Avoid steep slopes, pipe diameter changes, junctions, and areas of irregular shape due to breaks, repairs, root damage, debris, etc.
- Avoid locations that experience surcharging over the normal range of precipitation. Also, avoid sites that may experience backwater or tidal influences.
- If an automated sampler with a peristaltic pump is to be used, and the access point is a manhole, the water surface elevation should not be excessively deep.

Monitoring in Open Channels

- Take samples in locations of the reach that are stable and approach uniform flow. Sample sufficiently downstream of inflows to achieve a well-mixed condition across the channel and allow for "uniform" flow conditions. Saunders (1983) has guidelines on determining if the location is "well mixed."
- Stormwater is highly heterogeneous and the various particle sizes transported are not equally distributed from the bed to the surface of the water. Suspended particles less than about 0.04 mm are typically well mixed within the water column profile (Butler et al., 1996). As particle-size distribution (PSD) increases to include sand-size material (larger than 0.062 mm median diameter), a vertical gradient may form, with largest particles concentrating near the bed. Often it is necessary to integrate many samples (vertically or horizontally) to get a better representation of the water sample. In fluvial systems, a depth-integrated sample may be required because of potential variations in the cross sectional distribution of sediment (Guy, 1970).
- If depth integrated samples are not used, the sample should be collected in the middle portion of the water column where the suspended solids are average concentrations.

Monitoring BMPs

- The location of sampling should be located near the inlet and outlet of the BMP control structure to determine the removal provided by the BMP. Upstream monitoring stations should be located far enough away from the BMP to ensure that samples are independent of the BMP. Likewise, monitoring stations should be located immediately downstream so that BMP effluent is sampled before it is introduced into the receiving waters or is exposed to factors that may affect constituent concentrations
- Additional information should be reported when monitoring BMPs including type of BMP, BMP surface area, design capacity, inlet features, presence of overflow structures and characteristics. In addition, a description of the types and designs of outlets, localized channel type and geomorphology, slopes, stream systems, hydraulic response of BMP to hydrologic inputs, and location of separate inflow and outflow points. Finally, it is important to note installation data, any retrofitting, maintenance activities and upstream or downstream site characteristics.

Monitoring Flow and Precipitation

- Accurate measurement of the intensity and duration of each precipitation event and resultant total storm discharge is important to quantify the pollutant mass balance and effects upon a receiving water body (Thoman and Mueller, 1987; Irish et al., 1996).
- It is necessary to have accurate flow and rainfall measurements. It is recommended to use primary flow devices/methods (flumes, weirs) where possible, although volume-based methods, velocity-based methods, empirical equations, and tracer-dilution methods also

provide flow measurements. It is recommended to select a flow measurement method and location of measurements that will give accurate flow data to meet the objectives. Rapid changes in flow may be associated with rapid changes in the sediment concentration, PSD, and density distribution.

- Precipitation monitoring is important when studying urban runoff. Rainfall gauges should be located near sampling locations. Generally, the higher the number of precipitation gauges, the better the estimate of precipitation amounts.

The location of the samples should be in a well mixed area. In some studies, the concentrations of samples collected with an automatic pumping sampler were similar to those that were cross sectionally integrated (Krug and Goddard, 1986; Bossong et al., 2006). Therefore, if the location of sampling is judiciously placed, time intensive depth and width integration may not be necessary. Also, urban runoff often produces turbulent flows and rapid mixing characteristics associated with a well mixed sample. Frequently in urban drainage, depth and width integrated sampling may not be possible because of constraints such as personal and money constraints, and sampling issues including brief duration of runoff, limited access to the drainage structure, size of the conduit, depths and velocities of water in the conduit, and rapidly varying flows (Bent et al., 2001).

For all collection efforts, sampling devices must be made of chemical resistant materials that will not affect the quality of the stormwater sample. In general, Teflon, glass, or polyethylene is suitable sampling material.

It is important to validate and evaluate uncertainties and the representativeness of the samples. Experiments show that despite controlled and careful experimental conditions, relative uncertainties are about 20% for flow rates in sewer pipes, 6-10% for volumes, 25-35% for TSS concentration and loads, and 180-276% of TSS removal rate (Bertrand-Krajewski et al., 2003). Error uncertainties can be attributed to online sensors, location of sensors and samplers, sampling devices, and sampling time and space scales.

9.4.2 How to Take Samples

The best technique for obtaining samples depends on the researcher's situation. Samples can be collected manually or automatically. A discussion of manual, automated, grab and composite sampling is provided in Sections 4.2 and 4.3. In most cases, programmable automatic flow samplers with continuous flow measurements are recommended for most stormwater monitoring. Automatic samplers reduce additional variability induced by personal from sample to sample (Bailey, 1993). In addition, autosamplers can obtain data for many events and throughout the storm because they are programmable and do not require personnel in the field. They can initiate sampling very close to the beginning of runoff. Samples should be collected throughout the entire duration of the storm and an adequate number of samples should be taken to represent the solids concentrations in the runoff. Manual samples are recommended when an autosampler is not available and the sampling number is small. However, autosamplers are not capable of sampling large material (Gross solids), bedload or floating solids. Bedload samplers and nets designed to capture floatable material may be needed to supplement automatic samplers (Burton and Pitt, 2002). Manually-collected samples are required for constituents that transform rapidly. Advantages and disadvantages for composite and grab samples are discussed in Section 5.1.2.

Grab samples are the only option for monitoring parameters that transform rapidly (requiring preservation) or adhere to containers. Temperature, pH, total residual chlorine, phenols, volatile organic compounds, and bacteria transform rapidly therefore grab samples are appropriate if testing for these constituents. Oil and grease and TPH adhere to sample containers, therefore transferring samples between containers should be minimized. Grab samples show a snapshot of the water quality and are typically not sufficient to develop estimates of event mean pollutant concentrations unless a sufficient number are taken to represent the concentration changes over the period of runoff, and flow measurements are taken synoptically with the grab samples. The monitoring program should choose equipment that matches the requirements and goals of the sampling and monitoring program.

9.4.3 Handling and Holding Time

Sampling and holding equipment should be constructed of inert, nonreactive material. The U.S. EPA recommends specific materials for sample containers (plastic, glass, Teflon) for different constituents. It is important to consider this when choosing monitoring parameters. Some parameters require specific chemical preservation and different maximum holding times. Federal Register 40 CFR 136.3 lists recommended sample containers, preservatives, and maximum recommended holding times for constituents.

The various parameters have storage and holding times that are designated by the U.S. EPA. It is necessary to analyze the water sample as quickly as possible after sampling but in no case exceed the maximum holding time stated by the U.S. EPA. The standards for sample preservation and holding times are *ASTM D4841- 88/D4515-85/D3694-96.* The US EPA requires the solids samples to be stored at 4°C without freezing. The samples should be cooled to 4°C prior to transporting and maintained during transport to the laboratory. Reduce sample manipulation whenever possible prior to analysis. Settling characteristics tend to improve with increasing storage time, suggesting agglomeration of small particles (Dalrymple et al., 1975)

Some analysis on stormwater samples should be done as soon as possible or on site. Refer to the newest version of Standard Methods for constituents and their recommended holding times. It is recommended to analyze stormwater solids as soon as possible, but in no case hold for longer than seven days. Particles in stormwater runoff naturally grow and change causing significant changes in PSD over time periods of hours (Kayhanian, 2006). If particle size distribution is a concern, Kayhanian (2006) recommends that samples be analyzed within six hours of collection.

9.5 Gross Solids Sampling (Solids greater than 5 mm)

Gross solids are often neglected in stormwater sampling because the equipment used often has an inlet that is restricted to collecting material smaller than the nozzle. Sampling gross solids is difficult because some solids are buoyant while others tend to settle. The most buoyant types of floatable debris are plastics and some types of rubber. Paper, wood, leaves, and cloth items initially float but tend to sink once they become saturated with water. Glass, metal, and some types of rubber sink unless air is trapped in pockets of the material. Laboratory testing of gross pollutants showed that typically only 20% of the litter and less than 10% of the vegetation usually floats (Allison et al., 1998).

Gross solids are highly variable temporally and spatially. In order to normalize these variations, yearly data accumulation measurements of gross pollutants will provide more useful results than shorter time frequency comparisons. If the study is designed to compare characteristics of a specific rainfall event to gross solids accumulation, the captured material should be retrieved and analyzed at end of the storm.

Most trash and debris monitoring efforts have utilized discrete outfalls fitted with a trapping system, or catch-basin sumps located in strategic drainage basins with well defined land uses and drainage characteristics. If there is not a permanent BMP in place to capture gross solids, such as a gross solids trap, catch basin inserts, baffle boxes, or a hydrodynamic separator that includes capture capability for neutral and buoyant debris, nets should be used to sample gross solids. If practical, sampling gross solids should include the entire flow. This can be done at storm sewer inlets, outlets, or across an open channel. The gross solids analysis should be performed at the time of the cleanout. Estimate the volume of gross solids while still in the BMP, vacuum truck or after it is dumped at the disposal site. Some litter is floatable and can be skimmed off the top of the catchment device. The volume of the litter should be recorded at the time of collection. If the monitoring plan requires further characterization of the litter this should be recorded.

Many of the gross solids collections will require sub-sampling. If subsampling is necessary, the subsample should be representative of the solids captured (this may require a large sample). It should be taken away from the sides of the sampler to avoid potential contamination.

9.5.1 Gross Solids Handling

Rushton and England (Draft Report, 2006) recommend that gross solids samples be transported to the lab, or field work area, and processed as soon as possible with a holding time of no longer than 72 hours. If an anaerobic condition needs to be maintained, the solids should be transported and stored in an oxygen free apparatus. If volatile compounds are present the containers should be filled to the brim to reduce oxygen exposure. The samples should be kept at 4°C or frozen until analysis. If a sample is to be frozen, the container should be filled approximately 90% to allow for expansion. However, depending on the composition of the gross solids sample, which might include fairly large pieces of debris, or the size of the sample collected, parts of this protocol might be difficult to implement.

9.6 Recording Results

One of the reasons that it is difficult to acquire reliable and consistent results is the lack of detailed record keeping and reporting. It is difficult to compare data collected in various research efforts because of differences in sampling collection, handling, and methods of analyzing that are not well documented. When reporting stormwater monitoring results, the following information is recommended to be recorded:

- General monitoring information and goals
- Watershed information – watershed area, percent impervious, land use, vegetation, point source discharges, BMPs in watershed
- Sample location information - intake orientation, horizontal and vertical placement of sampling equipment
- Sample equipment data – type of sample, equipment calibration, equipment cleaning, intake diameter, method of collection, method for composite (if needed), equipment used for splitting

- Precipitation data – method for obtaining precipitation data, location of monitoring, antecedent dry period, unusual weather patterns that may affect the solids in the runoff
- Flow data – baseflow data, runoff, hydrograph data, etc.
- Water quality analytical data – holding time, detailed analytical method, filters size used, methods of mixing, pipettes used, etc.

Accurate reporting is one of the most important components of a monitoring program. Detailed reports will aid in improving monitoring and research efforts.

CHAPTER 10.0

Regulatory Issues Related to Substituting SSC for TSS Reduction Requirements

10.1 Introduction

Two main issues confronting regulatory agencies faced with changing laboratory analysis protocols and definitions of solids are: 1) deciding what standards are appropriate for stormwater discharges, and 2) how to assess the performance of various proprietary and public domain BMPs.

Many jurisdictions have adopted criteria for expressing the "quality" of stormwater and the effectiveness of their control programs in terms of TSS. Some stormwater programs require stormwater controls to reduce TSS by a given percentage. Control methods and BMPs are commonly described and rated in terms of their ability to remove a specified percentage of TSS or the allowable discharge of a specified concentration of TSS in the discharge. Such approaches may be appropriate when the relationships between TSS, pollution potential, and impacts on beneficial uses are well understood. However, because of the current diversity of sampling, analysis, and reporting practices, and the lack of a common "language" for expressing what is found by such monitoring efforts, it is likely that such relationships are not well understood.

Further complicating the problem is the use of SSC suggested by USGS instead of TSS as the more representative parameter for solids in stormwater. How would changes in definitions of stormwater-borne solids affect the various regulatory performance requirements in use today (e.g., *80% removal of TSS*)? Will these standards of performance continue to achieve the desired pollutant removals as were intended when TSS was specified or would another performance standard be more appropriate if SSC was substituted for TSS? The following sections discuss these issues.

10.2 BMP Performance Standards

As mentioned previously, many jurisdictions (e.g., Washington, Wisconsin, Maryland, New Jersey, Texas, etc.), have adopted a requirement of 80% TSS removal from runoff from new development. In many cases, TSS is not a primary concern for water quality, but is used as a surrogate for other contaminants that may potentially be in stormwater runoff. TSS was selected as a common, easy to analyze component. In addition, TSS concentrations may be correlated with those of other constituents, particularly those in or associated with a solid phase.

In a perfect world, agencies would assess each receiving water, determine its susceptibility to stormwater pollution, and evaluate its assimilative capacity for TSS and other constituents. This process is similar to that which occurs in the development of total maximum daily loads (TMDLs). However, because this assessment would be very time consuming and subsequent rules difficult to administer, most jurisdictions have adopted a single standard for stormwater discharges. Consequently, the 80% TSS removal efficiency commonly adopted does not reflect the site specific needs of the receiving water, or the need to limit the total mass of fine and coarse solids discharged to a certain level. Therefore, the fact that TSS measurements may understate the total mass of suspended solids discharged in runoff, is not a critical deficiency.

It is difficult to track the history of the use of the 80% standard back to the agency that first implemented it. In the current environment, many jurisdictions that are considering adopting a performance standard for BMPs default to the 80% value because it is in widespread use and appears that it can be achieved with a variety of public domain stormwater treatment facilities that are suitable for a wide range of climatic and site specific constraints. These include, but are not limited to: retention (wet) ponds, filtration systems, infiltration facilities, and some vegetated controls. Functionally, adopting an 80% TSS standard is equivalent to eliminating lower performing systems such as swales, filter strips, dry extended detention basins, and many proprietary controls (as frequently sized). Even though TSS measurements likely underestimate the total solids removal in BMPs, it does provide a relative ranking of solids removal performance that likely would not change if SSC procedures were used in BMP performance studies.

One might also ask whether the regulatory objective of 80% TSS removal is meant to be equivalent to 80% suspended solids removal. There are reasons to think that this might not be the case. Since the specific surface area is relatively greater for smaller particles, there is more opportunity for the adsorption of hydrophobic pollutants and even though the total mass of adsorbed pollutants may not be the largest (if the PSD is skewed towards bigger particles), the concentration will be highest. The relatively higher specific surface area results in higher concentrations of metals, PAHs, pesticides, and other pollutants in fine-grained sediment deposits that accumulate in receiving waters. It is also the smaller particles that have more adverse impacts on the benthic community. In many cases the larger material (sand and gravel) in runoff is desired to form a stable substrate in aquatic habitats. Consequently, it can be argued that regulatory agencies may be more interested in reducing the discharge of smaller particles, whose mass is more closely related to TSS concentration than SSC.

Another difficulty associated with changing BMP performance standards to SSC from TSS is that TSS removal for various types of BMPs is available from a variety of sources, including the International BMP Database, local monitoring efforts, and numerous published studies. On the other hand, there has been little or no monitoring of BMPs that use SSC for determining the suspended solids concentration. Consequently, if regulatory agencies adopted a removal percentage for SSC as a standard there would be no way to know which BMPs provided sufficiently high performance.

For the reasons described above, it would not be prudent at this time to replace a TSS removal threshold with SSC. As additional data become available regarding SSC concentrations in stormwater runoff and BMP performance, it may be possible in the future to make this change. Without a considerable research effort it could take as much as 20 years or so (about the same length of time TSS has been used in characterizing stormwater) before there will be sufficient data to support this potential change. At that time, there will be a much better understanding of what level of SSC removal might be required to achieve the same pollutant removal that is observed when removal of 80% of the TSS occurs.

Many authors have criticized the use of percentage removal of a constituent to characterize the pollutant removal performance of a BMP (Strecker et al., 2001; Barrett, 2003), and recommend using an approach that describes the effluent quality achieved. This is because calculated percent removal is strongly dependent on the influent concentration. BMPs located in watersheds with relatively clean runoff generally appear to be less effective than those in dirtier watersheds, even though the discharge quality of the two facilities may be identical. An advantage of characterizing performance by discharge quality is that in virtually all facilities, the coarse solids (fine sand and larger) will have been removed. When this is the case, both the TSS and SSC methodologies will produce essentially the same result. Based on an analysis of the BMP database by Lampe and others (2005) a discharge TSS or SSC concentration of about 25 mg/L would be the threshold that would discriminate between those BMPs that are currently considered to achieve the 80% threshold and those that do not.

10.3 BMP Evaluations

BMP performance evaluation is one area that has received a substantial amount of attention in the last few years. This attention is often instigated by manufacturers of proprietary stormwater treatment systems because of the importance of achieving the 80% TSS removal required in many jurisdictions. Their criticisms are directed at both how samples are collected and the type of analyses conducted in the laboratory, with the primary focus on the characterization of the influent quality. They believe (rightly so) that the use of automatic samplers potentially biases the samples since large particles may be excluded by the intake strainer or be too heavy for the peristaltic pump to lift to the sample container. In addition, since most studies use TSS rather than SSC, laboratory results are further biased, potentially resulting in a considerable underestimation of the influent sediment load. Conversely, effluent samples are generally considered valid since the sand sized material is mostly retained in the treatment system. Consequently, the monitoring may substantially understate the actual sediment removal efficiency.

The question that arises is whether BMP monitoring for selected facilities that use SSC to characterize influent concentrations is consistent with the regulatory requirement related to TSS removal. As described in the previous section, regulators may be more focused on reducing the discharge of small, highly polluted particles, rather than the total mass of all particles. In which case, the use of TSS would be more appropriate for characterizing influent concentrations and BMP removal efficiency. If one accepts that the 80% standard was selected solely on the basis of allowing only "better" performing BMPs to be implemented, then it must be remembered that all the BMP performance data that was used to develop this standard was based on the TSS procedure. This means that if BMPs are designed for 80% removal of SSC, both proprietary and public domain BMPs will under perform with respect to removal of TSS, the target parameter because of the "80%" has a higher fraction of heavy solids in the SSC measurement than in the TSS measurement.

An important consideration in developing BMP performance protocols is that they be equally applicable to public domain as well as proprietary BMPs. Most manufacturers of proprietary devices push for mass balance approach to eliminate the issues related to influent samplers not collecting the large material (Lippner et al., 2004a, b). It is generally assumed that the effluent sample is not biased, since all the large material will be captured in the device. A critical assessment must be made on whether this is a workable solution that could apply to all BMPs. The mass balance certainly provides a more accurate estimate of solids removal for proprietary devices that have small, well defined sumps. This is much more difficult to achieve in many public domain controls that in many cases are much larger and may have been constructed with earthen bottoms. For instance, determining the volume of material retained by a relatively small device seems fine, but is there a way to easily do this in a large public domain control like a retention (wet) pond? Would it realistically quantify the retained solids that might be widely dispersed on the bottom of a wet pond in a thin layer? There is also a question of whether we actually need to quantify the absolute solids removal performance in a public domain or proprietary treatment control. Perhaps it is sufficient merely to characterize the relative performance of these controls, so that we can choose the best product whether or not the total solids removal is known. If we only care about relative performance, then this can likely be obtained from an analysis of TSS alone.

Consequently, at the present time, there exist regulatory and BMP evaluation programs that are internally consistent in that they all are based on the use of the TSS laboratory procedure. Nevertheless, the fact that neither the rules nor most of stormwater monitoring data accurately characterizes the mass of the full range of particle sizes in stormwater is a concern. This may be a particular problem when maintenance frequencies for removal of accumulated sediment in BMPs is estimated based on TSS data. The question we face is whether to adopt, gradually and consistently, methodologies that provide more accurate measurements of the total mass of stormwater solids for regulatory related purposes. Immediate adoption of SSC for watershed studies is certainly appropriate, especially when the interest is related to sediment transport and accumulation (such as in reservoirs or channels). However, changing standards for administrative purposes seems less an immediate need, and certainly would have to include measurements of both SSC and TSS for some time to provide the data to adequately convert from one standard to the other. In addition, monitoring of both public domain and proprietary BMPs with the new standard would be necessary to define the performance of all BMPs using the same protocols.

Evaluating the gross solids removal of BMPs also remains an issue, which is especially difficult in public domain BMPs. As mentioned previously there are a variety of proprietary gross solids collection devices; however, it is important to remember that many public domain facilities also remove much or all of the gross solids fractions. Many BMPs have trash screens to prevent outlet clogging, which retain the gross solids, or devices such as sand filters permit only the discharge of particles smaller than the pores of the media. Determination of the amount of material collected is facilitated in proprietary BMPs by their relatively small size. This is much more problematic in large public domain systems, where the trash and debris that make up most of the gross solids component can be distributed over a wide area. Even where the amount of material collected can be quantified, most monitoring programs are not setup to determine what fraction either is not captured or bypasses the system during high flows. Since most gross solids removal is the result of physical straining, perhaps a study could be conducted to look at the affect of screen size opening on removal. This could be done by installing several screens in sequence with successively smaller openings. Then the gross solids removal performance of BMPs could be estimated simply from their design.

10.4 Research Needs

There is a pressing need for additional research related to the potential conversion from TSS to SSC protocols by regulatory agencies. These needs are related to three things: receiving water impacts, stormwater characterization, and BMP performance.

Additional information on the environmental impacts of coarse solids that are characterized more accurately by the SSC methodology on receiving waters is needed to determine whether the urban runoff regulatory focus needs to include the larger size fraction included in SSC. How does the presence of the larger size fraction in runoff impact channel morphology, especially the development of the pool/riffle structure and what is the effect on substrate for fish and benthic invertebrates. Another question regards the bioavailability of metals and other pollutants associated with the larger size fraction.

The widely cited study by Gray et al. (2000) that documented the bias of the TSS procedure for determining suspended solids mass considered only studies of natural waters. It would be extremely helpful to develop similar information regarding stormwater runoff properties and whether the bias was more or less than that observed in natural waters. This could be accomplished in at least two ways. A site that had been used for stormwater characterization studies could be reactivated with the monitoring focused on SSC. A comparison could then be made between TSS and SSC event mean concentrations from the two periods. Alternatively, monitoring could be conducted at a site with all samples analyzed for both TSS and SSC. This type of monitoring would need to be undertaken at numerous locations nationwide to account for the extreme variability in stormwater quality documented in previous studies.

To determine what SSC removal efficiency is equivalent to the 80% removal commonly specified for new development, additional monitoring of BMPs needs to be conducted using both the SSC and TSS protocols to facilitate comparison. It might be possible to combine this work with the characterization data needs described in the previous paragraph, where the BMP influent would characterize the runoff from the contributing watershed.

CHAPTER 11.0

SUMMARY AND RECOMMENDATIONS

11.1 Summary

Stormwater-born solids potentially degrade water quality, ecologic habitat, aquatic plants and animals, and may cause direct harm to human health. While the potential impact of various stormwater-borne solids is recognized and regulatory action is being taken, progress is hindered by the lack of common definitions or standardized monitoring procedures. This document has summarized the current state of stormwater solids characterization and sampling techniques. In order to improve current stormwater management practices, this report can aid in developing consistent definitions for characterizing, sampling, and analyzing stormwater solids.

Research efforts and stormwater regulations have measured solids in stormwater using various definitions and analytical protocols. It is proposed in this report that a classification system for stormwater solids be adopted based on particle size. This classification is divided into the following four categories: *Gross solids* defined as solids greater than 5 mm, *Coarse solids* defined as solids less than 5 mm and greater than 75 µm, *Fine solids* defined as solids less than 75 µm and greater than 2 µm, and D*issolved solids* defined as solids less than 2 µm in diameter. This classification offers a standard definition based on size which will aid in stormwater monitoring and research efforts. These definitions are based on operational definitions, sampling methods, and environmental impacts. The solids classification system provides for further subdivision within each of the four solids classed to include settleable or non-settleable fractions and volatile or non-volatile fractions based on analytical methods and needs for local water quality management planning.

Current analytical procedures for suspended solids determination cause uncertainty in the reported values for TSS because of loosely defined filter sizes, mixing techniques, and subsampling procedures. *It is recommended that the current Standard Methods protocol be revised into a well-defined procedure in order to standardize results of the TSS method. The recommended revision is to stir the sample at 600 rpm for 1 minute followed by a short settling time (1-5 minutes) and then take a subsample from the middle of the sample in the container.* The sample is then filtered over a 2 µm glass filter and analyzed in the same manner outlined in the standard methods procedure for TSS (SM 2540 D). Alternatively, the TSS analysis protocol could be modified to include the use of a separatory funnel to provide for separation and analysis of settleable solids and suspended solids. At a minimum, the procedure should be modified to require the sample to be stirred at 600 rpm for one minute before taking any subsamples and that a wide bore pipette be used to subsample from the mid-depth of the sample, midway between the side of the container and the vortex. The preferred alternative

A detailed plan for stormwater monitoring is necessary prior to any sampling. Obtaining a representative sample of stormwater proves to be a difficult task because stormwater solids concentration is variable both temporally and spatially. Therefore, it is necessary to first identify the goals and the solids of concern when developing a sampling plan. If gross solids are targeted, it is recommended to analyze gross solids using a gross solids removal device (such as a hydro separtator) or by installing nets or screens with 5 mm openings over outfall pipes or across surface channels. Coarse solids require a combination of autosamplers and bedload samplers. Fine solids and dissolved solids are transported primarily as suspended solids and the size less than 75 μm is sampled by most autosamplers. Therefore, autosamplers should be used to monitor fine and dissolved solids. There are several documents that outline stormwater monitoring plans including when to sample, equipment, sampling location, and number of samples (such as the Urban Stormwater BMP Performance Monitoring document). These documents should be consulted and a customized plan developed to effectively monitor stormwater.

11.2 Future Research and Recommendations

The accuracy of suspended solids measurements in urban runoff is dependent on a number of factors, including instream spatial and temporal variability, computational time frame, the ability to obtain a representative sample, proper deployment of appropriate samplers, appropriate sampling procedure, reliable shipping procedure, and the use of quality-assured analytical techniques by the laboratory (USGS 1998 in Gray, 2002). Further research is required into obtaining a representative sample of solids over a wide range of particle sizes. Autosamplers are limited by the size and location of the intake nozzle which severely restricts their ability to sample coarse and gross solids, thus these categories of solids are often neglected in stormwater monitoring programs. It is recommended that future research be focused on improving sampling equipment and methods to take into account the solids that are floating, transported in suspension, transported near bedload, and bedload. There is a need for better field equipment and insitu equipment to represent solids transported in urban runoff.

TSS is designed to characterize only one class of solid material commonly found in stormwater, yet it is commonly used to describe overall stormwater quality. Further investigation is required to improve existing protocols to include particle settling ability and PSD in order to advance the understanding of the transport, fate, and treatability of stormwater solids. Two modified protocols for TSS analysis are proposed in this report, both of which include settling time in the analytical methods to better separate the category of heavy solids from light solids providing more information on how the solids in stormwater will act in the environment. This document describes draft protocols for sample collection and analysis based on the four solids classifications defined within. Further investigations to validate the proposed analytical modifications that allow for a settling time are recommended. Efforts should be made to experiment with the recommended revised TSS analysis procedure, evaluating its efficacy as a more rigid and meaningful standardized procedure for classifying stormwater-borne solids.

Allowing a settling time before performing the TSS analysis would allow some particles to settle out, therefore making it more difficult to reach percent removal performance mandates (such as 80% TSS removal), however, accounting for settleability would be more representative of actual impact on human and ecosystem health. It is not all TSS that are negatively impacting receiving waters. Usually the fine solids are readily transported and are associated with a higher pollutant load. Larger particles may not be a problem, not where fines are readily transported away. Effluent concentrations may be a better indicator of BMP performance because percent removal is a function of influent concentration. It is therefore recommended to take the proposed solids definitions and align these with treatment methods and BMP performances. Effluent concentrations are better indicators of water quality impacts as opposed to BMP percent removal values. For example, heavily polluted influents may result in a higher percent removal as compared to influents with very minimal pollution concentrations, even though the effluent quality is significantly poorer. Considering this, site specific parameters should be defined and analyzed from the standpoint of target effluent concentrations obtained after treatment.

Identifying the solids that are transported in stormwater is the first step in effectively treating the solids using BMPs. Additional research on particle size distribution and the pollutant load distribution across the different particle size fraction is important in determining the critical solids that require treatment. Research on PSD and pollutants will aide in BMP selection based on methods to treat, remove, and immobilize stormwater solids. A comprehensive study on treatment of a range of particle size and associated pollutants linked to BMPs is critical in improving existing stormwater management.

Responsibilities of stormwater management to meet state and federal permitting are ever increasing. There is a need for information for stormwater managers and planners to better select and design BMPs based on improving impacts to water quality and comply with current and future TMDLs. TMDLs rely on BMPs to meet allocations by controlling stormwater solids. BMPs have been successful in improving receiving water quality, although there is not a comprehensive BMP program that is successful in meeting standards in all flow regimes. Research to link stormwater BMPs and the effectiveness of treating various particle sizes and controlling pollutant fate is a critical need given the increasing regulations being placed on stormwater.

REFERENCES

Allan, D.J. 1995. Stream Ecology. Structure and function of running waters. Kluwer Academic Publishers. p 324.

Allison, R.A., T.H.F. Wong, and T.A. McManon. 1996. Field trials of the polluted stormwater pollutant trap. *Journal of Australian Water and Wastewater Association*. September. p 29-34

Allison, R.A., T.A. Walker, F.H.S. Chiew, I.C. O'Neill, and T.A. McMahon. 1998. From roads to rivers: gross pollutant removal from urban waterways. Cooperative Research Center for Catchment Hydrology. Report 98/6.Department of Civil Engineering, Monash University, Clayton, Victoria 3800, Australia. www.catchment.crc.org.au/

American Society of Civil Engineers (ASCE) and Water Environment Federation (WEF). 1992. *Design and Construction of Urban Stormwater Management Systems*. ASCE, Manuals and Report of Engineering Practice No. 77, New York. N.Y. WEF, Manual of Practice No. FD-20, Alexandria, VA.

American Society of Civil Engineers/Environmental Protection Agency. 2002. A guidance manual for Best Management Practices (BMP) monitoring. U.S. EPA -821-B-02-001. www.bmpdatabase.org/docs.html

APHA, AWWA, WPCF. 1998. Standard Methods for the Examination of Water and Wastewater. 20th Edition. American Public Health Association, American Water Works Association, Water Pollution Control Federation, Washington, D.C.

Armitage, N. and A. Rooseboom, 2000a. The Removal of Urban Litter from Stormwater Conduits and Streams: Paper 1 The Quantities Involved and Catchment Litter Management Options. *Water South Africa*. 26, 181.

Armitage, N. and A. Rooseboom, 2000b. The Removal of Urban Litter from Stormwater Conduits and Streams: Paper 2 Model Studies of Potential Trapping Structures. *Water SA*. 26, 189.

Ashley, R.M., J.L. Bertrand-Krajewski, T.Hvitved-Jacobsen, and M. Verbanck. 2004. Solids in Sewers – Characteristics, Effects and Control of Sewer Solids and Associated Pollutants. *International Water Association Scientific and Technical Report No. 14* IWA publishing ISBN 1900222914

Ashley, R., J.L Bertrand-Krajewski, and T. Hvitved-Jacobsen. 2005. Sewer Solids – 20 years of investigation. *Water Science and Technology*. IWA Publishing. Vol 52 No 3 pp 73-84.

Bailey, R. 1993. Surviving the Stormwater Permit Process: *Water Environment & Technology*, vol. 5 no. 7 p. 70–74.

Barrett, M.E., 2003, "Performance, Cost and Maintenance Requirements of Austin Sand Filters, Journal of Water Resources Planning and Management, Vol. 129, No. 3, pp. 234-242.

Benda, L. and T. Dune. 1997. Stochastic forcing of sediment supply to channel networks from landsliding and debris flow. *Water Resources Research. 33(12):2849-2863.*

Bent, G.C., J.R. Gray, K.P. Smith, and G.D. Glysson. 2001. A Synopsis of Technical Issues for Monitoring Sediment in Highway and Urban Runoff. United States Geological Survey - File Report 00-497.

Bertrand-Krajewski, J.L., J.P. Bardin, M. Mourad, and Y. Béranger. 2003. Accounting for sensor calibration, data validation, measurement and sampling uncertainties in monitoring urban drainage systems. *Water Science and Technology* IWA Publishing. Vol 47 No 2 pp 95-102

Bledsoe, B.P. and C.C. Watson. 2001. Effects of urbanization on channel instability. *Journal of the American Water Resources Association* 37:255-270.

Blight, L.K. and A.E. Burger, 1997. Occurrence of plastic particles in seabirds from the eastern North Pacific. Marine Pollution Bulletin 3 (5): 323-325

Booth D.B. 1991. Urbanization and the natural drainage system – impacts, solutions and prognoses. *Northwest Environmental Journal* 7:93-118.

Booth, D. and L. Reinelt. 1993. Consequences of urbanization on aquatic systems – measured effects, degradation thresholds, and corrective strategies. Pages 545-550 *in* Proceedings Watershed '93 A national Conference on Watershed Management. March 21-24, 1993, U.S. Environmental Protection Agency, Alexandria, Virginia.

Booth, D. and C.R. Jackson. 1997. Urbanization of aquatic systems: degradation thresholds, stormwater detection, and the limits of mitigation. *Journal of American Water Resources Association* 33:1077-1090.

Bossong, C.R.; M.R. Stevens, J.T. Doerfer, and B.R. Glass. 2006. Summary and Evaluation of the Quality of Stormwater in Denver, Colorado, Water Years 1998-2001. United States Geological Survey. Scientific Investigations Report 2005-5150.

Bricker, O.P., B.F. Jones, and C.J. Bowser, 2003, Mass-balance approach to interpreting weathering reactions, *in* Drever, J.I., ed., Surface and Ground Water, Weathering and Soils, Treatise on Geochemistry, v. 5, chap 4: New York, Elsevier, p. 119-132.

Burton, G.A., Jr. and R.E. Pitt. 2002. Stormwater effects handbook: A toolbox for watershed managers, scientists, and engineers. Lewis Publishers: CRC Press, New York, NY.

Butler, D., R.W.O. May, and J.C. Ackers, 1996, Sediment transport in sewers Part 1—Background: Proceedings of the Institution of Civil Engineers-Water Maritime and Energy, v. 118, no. 2, p. 103–112.

Butler, D., J.W. Davies, C. Jefferies, and M. Schutze. 2002. Gross Solids Transport in Sewers. *Water & Maritime Engineering* 156 IssueWM2 p 175-183.

Californina Department of Transportation. 2000. Guidance for monitoring stormwater litter. Caltrans Doc. No. CTSW-RT-00-025.

California Department of Transportation. February 2002. *Trash TMDL Technical Report.*

California Department of Transportation Comprehensive Protocols Guidance Manual. 2003. California Department of Transportation. CTSW-RT-03-105.51.42

Center for Watershed Protection. 2003. Impacts of impervious cover on aquatic ecosystems. Center for Watershed Protection, Ellicott City, MD.

Chang, H.H. 1988. Fluvial Processes in River Engineering. New York, New York. John Wiley and Sons.

Chapra, S.C. and G.J. Pelletier. 2004. QUAL2K: A Modeling Framework for Simulating River and Stream Water Quality, Version 1.3: Documentation and Users Manual. Civil and Environmental Engineering Department, Tufts University, Medford, MA.

Chebbo, G. and A. Bachoc, 1992. Characterization of suspended solids in urban wet weather discharges. *Water Science and Technology* Vol 25 No 8 pp 171-179.

Chebbo G.; M.G. Gromaire; and M. Kafi-Benvahia. 2005. Spatial variability of characteristics and origins of urban wet weather pollution in combined sewers. *Water Science and Technology.* Vol 52 pp 53-62.

Cheesman B.C. and S.A. Williams,1999. Biodiversity and conservation of river macroinvertebrates on an expanding urban fringe: western Sydney, New South Wales, Australia. *Pacific Conservation Biology* Vol 5 pp 36-55.

Corbett C.W., M. Wahl, D.E. Porter, D. Edwards, and C. Moise 1997. Nonpoint source runoff modeling: A comparison of a forested watershed and an urban watershed on the South Carolina coast. *Journal of Experimental Marine Biology and Ecology* Vol 213 pp 133-149.

Cornelius, M.; T. Clayton, G. Lewis, G. Arnold, and J. Craig (1994). Litter Associated with Stormwater Discharge in Auckland City, New Zealand. Island Care New Zealand Trust, Auckland.

Crites, R.W. 1985. Micropollutant Removal In Rapid Infiltration. *Artificial Recharge Of Groundwater*. Butterworth Publishers, Boston. pp. 579-608.

Dallmer, L. 2002. SQIRTS- an On-site Stormwater Treatment and Reuse Approach to Sustainable Water Management in Sydney. *Water Science and Technology*, Vol 46 No 6 pp 151-158.

Dalrymple, R.J., S.L. Hodd, and D.C. Morin. 1975. Physical and Settling Characteristics of Particulates in Storm and Sanitary Wastewater. EPA-670/2-75-011 (NTIS PB 242 001), Cincinnati, OH – Edison NJ. U.S. Environmental Protection Agency.

Davies P.H. 1986. Toxicology and Chemistry of Metals in Urban Runoff. Urban Runoff Quality – Impact and Quality Enhancement Technology. P 60-78. (Eds B. Urbonas and L.A. Roesner) American Society of Civil Engineers.

Dodge, C. 2005. AMAFCA/Albuquerque MS4. Floatable and Gross Pollutant Study. http://www.cabq.gov/flood/pdf/FINALREPORT-OCTOBER2005.pdf.

Dodson, R.D. 1999. Stormwater Pollution Control. 2nd edition. McGraw-Hill Companies Inc. New York. Chapter 9.

Doll, B.A., D.E. Wis-Frederick, C.M. Bucknet, S.D. Wilkerson, W.A. Harman, R.E. Smith, and J. Spooner. 2002. Hydraulic geometry relationships for urban streams throughout the piedmont of North Carolina. *Journal of the American Water Resources Association* Vol 38 pp 641-651.

Driscoll, E.D., G.E., Palhegyi, E.W. Strecker, and P.E. Shelley. 1989. Analysis of Storm Events Characteristics for Selected Rainfall Gauges throughout the United States. United States Environmental Protection Agency. Washington, D.C.

Driscoll, E.D. 1989. Long Term Performance of Water Quality Ponds. In: *Design of Urban Runoff Quality Controls*. Edited by Larry a. Roesner, Ben Urbonas and Michael B. Sonnen. Published by the American Society of Civil Engineers.

Edwards, T.K. and G.D. Glysson. Field Methods for Measurement of Fluvial Sediment. 1999 USGS. Book 3, Apllications of Hydraulics, Chapter 2. Information Services, Box 25286, federal Center, Denver, CO 80225.

Ellis, J.B. and D.M. Revitt, 1982. Incidence of heavy metals in street surface sediments: solubility and grain size studies. *Water, Air, and Soil Pollution*. Volume 17, 87-100.

Federal Inter-Agency Sedimentation Project. 1963. Determination of fluvial sediment discharge: Minneapolis, Minnesota, St. Anthony Falls Hydraulics Laboratory, Inter-Agency Report 14, 151 p 60

Field, R., D. Averill, T.P. O'Conner, and P. Steel. 1997 Vortex Separation Technology, Water Quality Resources Journal, Canada, Volume 32,No 1

Garie, H.L. and A. McIntosh. 1986. Distribution of benthic Macroinvertebrates in a Streams Exposed to Urban Runoff. *Journal of American Water Resources Association* Vol 22 pp 447-445.

Gaugler, L. 2003. Floatables Action Plan Assessment Report http://www.epa.gov/Region2/water/action_plan/assess2003.pdf .

Glysson, G.D., J.R. Gray, and L.M. Conge, 2000, Adjustment of total suspended solids data for use in sediment studies: Proceedings, ASCE's 2000 Joint Conference on Water Resources Engineering and Water resources Planning and Management, July 31 – August 2, 2000, Minneapolis, Minn., 10 p. (http://water.usgs.gov/osw/pubs/ASCEGlysson.pdf).

Gordon, M. and R. Zamist. c. 2005. Municipal Best Management Practices for Controlling Trash and Debris in Stormwater Runoff. http://www.plasticdebris.org/Trash_BMPs_for_Munis.pdf

Gray, J.G; D. Glysson; L.M. Turcious; and G.E. Schwarz. 2000. Comparability of Suspended-Sediment Concentration and Total Suspended Solids Data. *United States Geological Survey* Water-Resources Investigation Report 00-4191.

Gray, J.R. 2002. The Need for Surrogate Technologies to Monitor Fluvial-Sediment Transport.http://water.usgs.gov/osw/techniques/TSS/listofabstracts.htm. Proceedings of the subcommittee on sedimentation's "Turbidity and Other Sediment Surrogates Workshop" 2002. Reno, NV.

Grout, H.; M.R. Wiesner, and J.Y. Bottero, 1999. Analysis of Colloidal Phases in Urban Stormwater Runoff. *Environmental Science and Technology* (G.B.) Vol. 33 No. 6 pp 831.

Gupta, G. and F.L. Harrison. 1981. Effects of Cations on Copper Adsorption by Kaolin. *Water, Air and Soil Pollution*. Vol. 15 pp 323-329.

Guy, H.P. 1969. *Laboratory Theory and Methods for Sediment Analysis,* Techniques of Water-Resources Investigations of the United States Geological Survey, Book 5, Chapter C1.

Guy, H.P. 1970. Sediment problems in Urban Areas. Water in the Urban Environment, United States Geological Survey Circular 601-E, Washington, D.C.

Hadley, R.F. and S.A. Schumm. 1961. Sediment sources and drainage-basin characteristics in Upper Cheyenne River Basin. United States Geological Survey. Water-Supply Paper 1531-B. 137-196.

Hammer, T.R. 1972. Stream channel enlargement due to urbanizations. *Water Resources Research* Vol 8 pp1530-1540.

Hannah, M. 2005. Stormwater Bed Load & Gross Pollutant Export Rates and Their Implication for Treatment Devices. In Fourth South Pacific Conference on Stormwater and Aquatic Resource Protection 4-6 May 2005, Aukland, NZ .

Harper, H. 1998. An explanation of sources of stormwater pollution and pollutant loading rates for different land uses in Florida. Environmental Research and Design, In. Orlando.

Hatt B.E., T.D. Fletcher, C.J. Walsh . and S.L. Taylor 2004. The influence of urban density and drainage infrastructure on the concentrations and loads of pollutants in small streams. *Environmental Management* Vol 34 pp 112-124.

Herricks, E.E. 2001. Observed Stream Responses to Changes in Runoff Quality. *In* Ben R. Urbonas eds. *Linking Stormwater BMP Designs and Performance to Receiving Water Impact Mitigation*. American Society of Civil Engineers. New York, NY p. 145-154.

Hijioka, Y.; H. Frurumai, and F. Nakajima, 2000. Runoff Pollutant Load of Suspended Solids with Different Particle Sizes in a Separate Sewer System. *Proc. 3rd JSWE Symposium*, September 2000, Neyagawa, Japan. Japanese Society of Water Environment (JSWE).

Horner, R.H., J.J. Skupien, E.H. Livingston, and H.E. Shaver. 1994. *Fundamentals of Urban Runoff Management: Technical and Institutional Issues*, Terrene Institute, Washington D.C., , p. 52–53.

Horowitz, A.J. 1995. The Use of suspended Sediment and Associated Trace Elements in Water Quality Studies. United States Geological Survey, Georgia District Office, 3039 Amwiler Road, Atlanta, Georgia. IAHS Special Publication No. 4.

Hubbell, D.W. and D.Q. Matejka, 1959, Progress report no. 1, Investigations of some sedimentation characteristics of a sand-bed stream: U.S. Geological Survey Open-File Report, 78 p.

HydroQual, Inc. NYPD2000. Floatables Pilot Program Final report: Evaluation of Non-Structural Methods to Control Combined and Storm Sewer Floatable Materials. Department of Environmental Protection, Division of Water Quality Improvement. Dec 1995.

Irish, L.B., W.G. Lesso, M.E. Barrart, J.F. Malina, R.J. Charbeneau, and G.H. Ward, 1996, An evaluation of the factors affecting the quality of highway runoff in the Austin, Texas area: U.S. Federal Highway Administration, Texas Department of Transportation, Interim Report, FHWA/TX-96/1943-5, 246 p.

James, R.B. 1999. Characteristics & Measurement of Solids in Stormwater Runoff. Paper presented at the Indiana Water Environment association.

James R.B. 2003. *TSS – a Viable Measure of Stormwater Pollutants?* http://www.stormwater-resources.com/Library/ From Stormcon 2003.

Jefferies C.J. and R.M. Ashley. 1994. The behaviour of gross solids in sewer systems. *Journal of the European Water Pollution Control Association,* Vol 4 No. 5, 11-17.

Jones-Lee, A. and G. Fred Lee. 2005. Unrecognized pollutants; US EPA Workshop on Pharmaceuticals in the Environment. Stormwater Runoff Water Quality Science/Engineering Newsletter Devoted to Urban/Rural Stormwater Runoff. Water Quality Management Issues. *EnviroQual* Vol 8. No 5 11 pages.

Karamalegos, A.M., M.E. Barrett; D.F. Lawler; J.F. Malina. 2005. Particle Size Distribution of Highway Runoff and Modification through Stormwater Treatment. CRWR Online Report 05-10. http://www.crwr.utexas.edu/reports/pdf/2005/rtp05-10.pdf

Kayhanian, M., T.M. Young; and M.K. Stenstrom. 2005. Limitations of current solids measurements in stormwater runoff. *Stormwater* July/Aug.

Kayhanian, M. J.E. Leatherbarrow, and C.P. Alaimo. 2006. Improved method of Suspended Solid Measurements. Presented at Stormcon conference 2006. Denver CO.

Kayhanian, M. 2006. On-line and Continuous flow-through particle size distribution measurement of stormwater runoff. Presented at Stormcon. 2006. Denver CO.

Kennen, J.G. 1999. Relation of macroinvertebrate community impairment to catchment characteristics in New Jersey streams. *Journal of American Water Resources Association* Vol 35 pp 939-955.

Kimble-Contes, Inc., accessed May 1, 2000, at URL http://www.kimble-knotes.com/html/pg-37005.html and URL http://www.kimble-knotes.com/html/pg- 37034B.html.

Kobriger, N.K. and A. Geinopolos,. 1984. Sources and Migration of Highway Runoff Pollutants. Research report, Vol. III. Rep. No. FHWA/RD-84/059, United States Department of Transportation, Federal Highway Administration, Washington D.C.

Krein, A. and M. Schorer. 2000. Road runoff pollution by polycyclic aromatic hydrocarbons and its contribution to river sediments. *Journal of American Water Resources Association.* Vol 34 No 16 pp. 4110-4115.

Krug, W.R. and G.L. Goddard, 1986, Effects of urbanization on streamflow, sediment loads, and chemical morphology in Pheasant Branch Basin near Middleton, Wisconsin: U.S. Geological Survey Water- Resources Investigations Report 85-4068, 82 p.

Lampe, L., M. Barrett, B. Woods-Ballard, A. Andrews, P. Martin, R. Kellagher, M. Hollon, and C. Jefferies. 2005, *Performance and Whole Life Costs of Best Management Practices and Sustainable Urban Drainage Systems: Final Report 2005*, Water Environment Research Foundation Project 01-CTS-21T, Alexandria, VA.

Landers, J. 2006. Selecting Stormwater BMPs. *Stormwater Journal*. SW May/June 2006. http://www.stormh2o.com/sw_0605_selecting.html

Lane, E.W. 1947. Report of the Subcommittee on Sediment Terminology, Transactions, American Geophysical Union, Vol. 28, No.6, Washington, D.C. pp 936-938.

Lane, S.L., S. Flanagan, and F.D. Wilde,. 2003. Selection of equipment for water sampling (ver. 2.0): United States Geological Survey Techniques of Water-Resources Investigations, book 9, chap. A2, accessed __*date*__ at http://pubs.water.usgs.gov/twri9A2/

Lenat D.R., D.L. Penrose and K.W. Eagleson. 1981. Variable effects of sediment addition on stream benthos. *Hydrobiologial*. Vol 79 pp.187-194.

Lenat, D.R. 1984. Agriculture and stream water quality, A biological evaluation of erosion control practices. *Environmetnal Management.* Vol 8 No. 4 pp. 333-344.

Lenhart, J. 2006. Affect of Variant Time to Sub-Sampling on Total Suspended Solids Results. Contech Stormwater Solutions Product evaluation. PE 38.09 Report

Lenhart, J., J Lehman. 2006. Measuring Total Volatile Suspended Solids in Stormwater to Understand the Influences of Organic Matter on BMP Performance. *Proceedings of the EWRI Conference. Omaha, NB, May, 2006.*

Lippner, G.; R. Churchwell, R. Allison, G. Moeller, and J. Johnston. 2000. A Scientific Approach to Evaluating Stormwater Best Management Practices for Litter. Presented at California Water Environment Association 72nd Annual Conference, Sacramento, CA 2000.

Lippner, G., A. Singh, and R. James, 2004, Evaluation of Monitoring and Analysis Methods Yielding Significant Differences in BMP Efficiency, Transportation Research Board Annual Meeting, January 11-15, 2004.

Lloyd, S.D., T.H.F. Wong, and B. Porter. 2001 "Implementing an ecological sustainable stormwater drainage system in a residential development" Water Science and Technology 45: 1-7.

Maxted, J. and E. Shaver. 1997. The use of retention basins to mitigate stormwater impacts on aquatic life. Pages 494-512 *in* L.A. Roesner, editor. Effects of watershed development and management on aquatic ecosystems, American Society of Civil Engineers, New York.

McCabe, J.M. and C.L. Sandretto. 1985. Some Aquatic Impacts of Sediment, Nutrients, and Pesticides in Agricultural Runoff. Publication No. 201. Limnological Research Laboratory, Dept. of Fisheries and Wildlife, Michigan State University.

Minton, G.R. 2005. *Stormwater Treatment: Biological, Chemical, and Engineering Principles.* Second edition.

Morquecho, RE. 2005. Pollutant Associations with Particulates in Stormwater Solids. Dissertation for a Doctor of Philosophy in the Department of Civil and Environmental Engineering. University of Alabama.

Mudgway, L.B., H.P. Duncan, T.A. McMahon, and F.H.S. Chiew. 1997. Best Practice Environmental Management Guidelines for Urban Stormwater, Cooperative Research Centre for Catchment Hydrology.

Murray, K.S., D. Cauvet, M. Lybeer, and J.C. Thomas, 1999. Particle Size and Chemical Control of Heavy Metals in Bed Sediment From the Rouge River, Southeast Michigan. Environment Science and Technology 33:987-992.

Muthukaruppan M., F. Chiew, T. Wong. 2002. Size Distribution and Partitioning of Urban Pollutants.

National Field Manual for Collecting Water Quality Data, 2005. USGS, Techniques of water resources investigation, Book 9 Handbook for water resources Investigation.

Natural Resources Defense Council. 1997. "Testing the Waters VII: How Does Your Vacation Beach Rate? Seventh Annual Report on Beach Pollution http://www.nrdc.org/nrdcpro/ttw/titinx.html.

New York – New Jersey Harbor Estuary Program Including the Bight Restoration Plan. 1996. Final Comprehensive Conservation and Management Plan.

Osborne L.L. and M.J. Wiley 1988. Empirical relationships between land-use/cover and stream water quality in an agricultural watershed. *Journal of Environmental Management* Vol 26 pp 9-27.

Paul, M.J. and J.L. Meyer. 2001. Streams in the urban landscape. Annual Review of Ecological Systems Vol 32 pp 333-365.

Pechacek, L.D. 1993. Urban Stormwater Runoff Characteristics Based on Singular land-use activity and Particle Size Distribution. Thesis – Department of Civil and Environmental Engineering. University of Houston.

Pisano, W.C. 1996. Summary: United States "Sewer Solids" Settling Characterization Methods, Results, Uses and Perspectives. *Water Science and Technology* Vol. 33 No. 9 pp 109-115.

Pitt, R. 1979 Demonstration of Nonpoint Pollution Abatement Through Improved Street Cleaning Practices. Untied States Environmental Protection Agency. Grant No. S-804432. EPA-600/2-79-161. 270 pages. Cincinnati, OH.

Pitt, R., S. Clark, and K. Parmer.1994. *Protection of Groundwater from Intentional and Nonintentional Stormwater Infiltration*. U.S. Environmental Protection Agency, EPA/600/SR-94/051. PB94-165354AS, Storm and Combined Sewer Program, Cincinnati, Ohio. 187 pgs.

Pitt, R., M. Lalor and J. Eston. 2001. Potential Human Health Effects Associated with Pathogens in Urban Wet Weather Flow. *Journal of American Water Resources Association*. http://unix.eng.ua.edu/~rpitt/Publications/MonitoringandStormwater/Stormwater%20Pathogens%20JAWRA.pdf#search=%22human%20health%20and%20stormwater%20bacteria%20pathogens%22

Pitt, R. 2001. Stormwater Management for Highway Runoff. *Symposium on the Pollution of Water Sources from Road Run-off*. Sponsored by the Committee for Public Transportation, the Faculty of Life Sciences, the Institute for Nature Conservation Research, and the Porter School of Environmental Studies at Tel Aviv University.

Pitt, R, M. Libum, S.R. Durrans, S. Burian, S. Nix, , J. Voorhees, and J Martinson. 2002. *Guidance Manual for Integrated Wet Weather Flow (WWF) Collection and Treatment Systems for Newly Urbanized Areas (New WWF Systems)* EPA/600

Radtke, D.B., Revised June 2005, Bottom-material samples: U.S. Geological Survey Techniques of Water-Resources Investigations, book 9, chap. A8, accessed __*Aug. 2006*__ at http://pubs.water.usgs.gov/twri9A8/.

Richardson, E.V., D.B. Simons, and P.F. Lagasse. 2001. River Engineering for Highway Encroachment. Highways in the River Environment. Hydraulic Design Series Number 6. Federal Highway Association NH1 01-004.

Rickert, D.A. 1997. Comparison of the Suspended-Sediment Splitting capabilities of the churn and cone splitters. Office of Water Quality Technical Memorandum No. 97.06.

Richter B.D., J.V. Baumgartner Wigington, and D.P. Braun. 1997. How much Water does a River Need? *Freshwater Biology*. Vol 37 pp 231-249.

Roberts, C.R. 1989. Flood frequency and urban-induced channel change: some British examples. Pages 57-82 *in* K. Beven and P.Carling, editors. Floods: hydrological, sedimentological and geomorphological implications. Wiley, New York.

Roesner L.A. 1999. The Hydrology of Urban Runoff Quality Management. *Sustaining Urban Water Resources in the 21st Century*. ASCE. Reston, VA. 229-241.

Roesner, L.A. and B.P. Bledsoe. 2003. Physical effects of wet weather flows on aquatic habitats: present knowledge and research needs. Water Environment Research Foundation, Report 00-WSM-4, Alexandria, VA.

Rushton, B., G. England. 2006. ASCE Guidelines for Monitoring Stormwater Gross Solids. Working paper for the Task Committee on Gross Solids, UWRRC, ASCE. 20 pages.

Sansalone, J.J. and S.C. Buchberger. 1997. Characterization of particulate loading in the urban stormwater first flush. *Water Science and Technology*, Vol 36 No. 9 pp. 155-160.

Sansalone, J.J., J. Koran, S.G. Buchberger, and J. Smithson. 1998. Physical characteristics of highway solids transported during rainfall. *ASCE Journal of Environmental Engineering* Vol. 124 No. 5 pp 427-440.

Sansalone, J. and T. Tribouillard. 1999. *Variation in characteristics of abraded roadway particles as a function of particle size – Implications for water quality and drainage.* Transportation Research Record 1690.

Sartor, J.D. and G.B. Boyd. 1972. *Water Pollution Aspects of Street Surface Contaminants,* EPA-R2-72-081.

Saunders, T.G. 1983. *Design of Networks for Monitoring Water Quality.* Water Resources Publications. December.

Shaheen, D.G. 1975. *Contributions of Urban Roadway Usage to Water Pollution* EPA 600/2-75-004.

Simons, D.B. and F. Senturk. 1991. Sediment Transport Technology. Water and Sediment Dynamics. Water Resources Publication, Littleton CO.

Snodgrass, W.J., B.W. Kilgour, M. Jones, J. Prish, and K. Reid. 1997. Can environmental impact of watershed development be measured In: Roesner, L.A. eds. *Effects of watershed development and management on aquatic ecosystems*. Proceedings of an Engineering Foundation Conference. Published by ASCE, New York 1997

Stahre, P. and B. Urbonas. 1990. Stormwater Detention for Drainage, Water Quality, and CSO Management. Prentice-Hall Inc. Englewood Cliffs, NJ.

Strecker, E.W., M.M. Quigley, B.R. Urbonas, J.E. Jones, and J.K. Clary. 2001. Determining Urban Storm Water BMP Effectiveness. *Journal Water Resources Planning and Management*, Vol. 127, No. 3, pp. 144-149.

Stenstrom, R.S., J.C. Clausen, and D.R. Askew. 2002. Treatment of Parking Lot Stormwater Using a StormTreat System. *Journal of Environmental Science and Technology*; Vol 36 No. 20 pp. 4441-4446.

Sullivan, J., F. McMasters, and T. Sobelman. 2005. LA River Cleanup. *Stormwater* online:http://www.gradingandexcavation.com/sw_0511_great.html

Thoman, R.V. and J.A. Mueller. 1987, Principles of surface water quality modeling and control: New York, Harper and Row, Publishers, 644 p.

Trimble, S.W. 1997. Contribution of stream channel erosion to sediment yield from an urbanizing watershed. *Science* Vol. 278 pp1442-1444

U.S. Army Corps of Engineers (USACE). 1995. Sediment Investigations in Rivers and Reservoirs. EM 1110-2-4000. Washington, D.C.

U.S. Environmental Protection Agency. 1983. *Results of National Urban Runoff Program (NURP).*

U.S. Environmental Protection Agency. 1992. Plastic Pellets in the Aquatic Environment: Sources and Recommendations. U.S. EPA 842-B-92-010. Washington, D.C.

U.S. Environmetnal Protection Agency. 1993. *Investigation of Inappropriate Entries into Storm Drainage Systems: A User's Guide*. EPA-600-R-92-238. U.S. Environmental Protection Agency, Office of Research and Development, Washington, D.C.

U.S. Environmental Protection Agency. 2000. National Water Quality Inventory. 1998 report to Congress U.W. Environmental Protection Agency.

U.S. Environmental Protection Agency, *Assessing and Monitoring Floatable Debris*. 2002a. EPA 842-B-02-002. Prepared by Oceans and costal Protection Division 2002.

U.S. Environmental Protection Agency 2002b. *Urban Stormwater BMP Performance Monitoring: A Guidance Manual for Meeting the National Stormwater BMP Database Requirements*. EPA 821-C-02-005. U.S. Environmental Protection Agency, Office of Water, Washington, D.C. April.

U.S. Environmental Protection Agency. 2003. excerpted from U.S. EPA 2003. "Developing Water Quality Criteria for Suspended and Bedded Sediments (SABS): Potential Approaches (Draft, August 2003)." U.S. EPA, Office of Water draft report. http://www.epa.gov/WARSSS/sedsource/sabs.htm

U.S. Geological Survey, variously dated, National field manual for the collection of water-quality data: U.S. Geological Survey Techniques of Water-Resources Investigations, book 9, chaps. A1-A9, available online at http://pubs.water.usgs.gov/twri9A.

U.S. Geological Survey. 1998. A national quality assurance program for sediment laboratories operated or used by the Water Resources Division: U.S. Geological Survey, Office of Surface Water Technical Memorandum No. 99.01 http://water.usgs.gove/admin/memo/SW/sw98.05.html.

U.S. Geological Survey Policy. 2000. Collection and Use of Total Suspended Solids Data, November 27, 2000 Offices of Water Quality and Surface Water Technical Memorandum.

Vignati, D.A.L., T. Dworak, B. Ferrari, B. Koukal, J. Loizeau, M. Minouflet, M.I. Camusso, S. Polesello, and J. Dominik, 2005. Assessment of the Geochemical Role of Colloids and Their Impact on Contaminant Toxicity in Freshwaters: An Example from the Lambro-Po System (Italy). *Journal of Environmental Science and Technology* Vol. 39 No. 2 pp. 489-497.

Walling D.E. 1983. The sediment delivery problem. *Journal of Hydrology*. Vol. 65 pp. 209-237.

Walsh, C.J., A.W. Leonard, A.R. Ladson, T.D., and Fletcher. 2004. Urban Stormwater and the Ecology of Streams. CRC for Freshwater Ecology and CRC for Catchment Hydrology: Canberra.

Walters, R.A. 1995. Chapter 1: Modeling surface water flow, in Carey, G.F., ed., Finite Element Modeling of Environmental Problems: Wiley and Sons, New York, p. 1-16.

Wang, L., J. Lyons, P. Kanehl, R. Bannerman, and E. Emmons. 2000. Watershed urbanization and changes in fish communities in southeastern Wisconsin streams. *Journal of American Water Resources Association* Vol. 36 pp. 1173-1189.

Wang, L., J. Lyons, P. Kanehl, and R. Bannerman. 2001. Impacts of urbanization on stream habitat and fish across multiple spatial scales. *Journal of Environmental Management* Vol. 28 pp. 255-266.

Welch E.B. 1992. Ecological Effects of Wastewater. (Cambridge University Press: Cambridge.) In. Walsh C.J., A.W. Leonard, A.R. Ladson, T.D. Fletcher Eds. *Urban Stormwater and Ecology of Streams*. CRC for Freshwater Hydrology 2004.

Welch E.B., J.M. Jacob, and C.W. May. 1998. Stream quality. In R. J. Naiman and R.E. Bilby eds. *River Ecology and management.* Lessons from the Pacific Coastal Ecoregion. 1999 pp 69-94. Springer NY.

Wilber, C.G. 1983. *Selenium: A Potential Environmental Poison and a Necessary Food Constituent.* Springfield: Charles C. Thomas Publisher.

Wilde, F.D. 2005, Preparations for water sampling: U.S. Geological Survey Techniques of Water-Resources Investigations, book 9, chap. A1; Urban Stormwater BMP Performance Monitoring: EPA-821-B-02-001 April 2002

Winterstein, T.A. and H.E. Stefan. 1986. Suspended sediment sampling in flowing water – Laboratory study of the effects of nozzle orientation, withdrawal rate and particle size: Minneapolis, St. Anthony Falls Hydraulic Laboratory External Memorandum M-168

Wolman, M.G. 1967. A cycle of sedimentation and erosion in urban river channels. Geografiska Annaler Vol. 49 pp. 385-395.

Woodward-Clyde Consultants 1999. Stormwater Inlet Insert Devices Literature Review. In. Edde, Wendy and Adam Gliveri eds. *Santa Clara Valley Urban Runoff Pollution Prevention Program*. 2002.

Wong T., P. Breen, and S. Lloyd. 2000. Water sensitive road design: design options for improving stormwater quality of road runoff. Cooperative Research Centre for Catchment Hydrology, Technical Report 00/1.

Yoder, C.O. and E.T. Rankin. 1997. Assessing the condition and status of aquatic life designated uses in urban and suburban watersheds. Pages 201-227 *in* L.A. Roesner, eds.. Effects of watershed development and management on aquatic ecosystems. American Society of Civil Engineers, New York.

IWA Publishing's authorised EU representative for General Product Safety Regulations is Diane D'Arras, 15 rue Duret, 75116 Paris, France, e-mail: safety@iwap.co.uk.

Printed and bound by CPI Group (UK) Ltd, Croydon, CR0 4YY
12/05/2026
02108868-0010